INORGANIC CONTROLLED RELEASE TECHNOLOGY

Materials and Concepts for Advanced Drug Formulation

INORGANIC CONTROLLED RELEASE TECHNOLOGY

Materials and Concepts for Advanced Drug Formulation

XIANG ZHANG

MARK CRESSWELL

ELSEVIER

AMSTERDAM • BOSTON • HEIDELBERG • LONDON
NEW YORK • OXFORD • PARIS • SAN DIEGO
SAN FRANCISCO • SINGAPORE • SYDNEY • TOKYO
Butterworth-Heinemann is an imprint of Elsevier

B
H

Butterworth Heinemann is an imprint of Elsevier
The Boulevard, Langford Lane, Kidlington, Oxford OX5 1GB, UK
225 Wyman Street, Waltham, MA 02451, USA

British Library Cataloguing in Publication Data
A catalogue record for this book is available from the British Library

Library of Congress Cataloging-in-Publication Data
A catalog record for this book is available from the Library of Congress

For information on all Butterworth-Heinemann publications visit our website at http://store.elsevier.com/

ISBN: 978-0-08-099991-3

CONTENTS

About the Author *vii*
Preface *ix*
Acknowledgments *xi*
Key Features *xiii*

1. Materials for Inorganic Controlled Release Technology **1**

1.1 Introduction 1
1.2 Comparison between Organic and Inorganic CRT 2
1.3 Materials Chemistry and Processing Technology 7
1.4 Materials Physics and Drug-Loaded Micro/Nanostructure 12
References 15

2. Materials Fundamentals of Drug Controlled Release **17**

2.1 Introduction of Materials Nanostructure 17
2.2 API Distribution Within Inorganic Matrices 23
2.3 Basic Understanding of Potential Molecular Interactions 29
2.4 Theory and Practical Modelling of Drug Controlled Release Kinetics 48
References 53
Further reading 55

3. Materials Characterization of Inorganic Controlled Release **57**

3.1 Introduction 58
3.2 Chemical Analysis 59
3.3 Physical Property Analysis 72
3.4 Microscopy 82
3.5 Technology for Evaluation of API and Inorganic CRT Carrying System 84
References 90
Further reading 91

4. Silica-Based Amorphous Drug Delivery Systems **93**

4.1 Introduction 93
4.2 Sol-Gel Chemistry and Processing 94
4.3 Ordered Mesoporous Silica 98
4.4 Non-Ordered Porous Silica for Controlled Release 120

4.5 Porous Silica for Improved Efficacy and Tolerability of Topically Administered Active Compounds 131
References 134

5. Mesoporous Bioactive Glass-Based Controlled Release Systems 139
5.1 Introduction 139
5.2 General Synthesis Methods for Mesoporous Bioactive Glasses 140
5.3 Mesoporous Glass-Based 3D Scaffold Structures 141
5.4 The Role of MBG Surface Functionalization for Controlling Active Release 146
5.5 Case Studies 148
5.6 Conclusions and Future Context 154
References 154

6. Calcium Phosphate Materials for Controlled Release Systems 161
6.1 Calcium Phosphate Materials and Their Properties 161
6.2 Calcium Phosphate Cements 164
6.3 Calcium Phosphate Foams for Drug Delivery 178
6.4 Drug Release from Calcium Phosphate Coatings 179
References 183

7. Alternative Inorganic Systems for Controlled Release Applications 189
7.1 Introduction 189
7.2 Clays 190
7.3 Alternative Mesoporous Inorganic CRT Materials 207
7.4 Nanoparticle Delivery Systems 211
References 216

8. Future Development of iCRT 221
8.1 Overview 222
8.2 Design and Development of Inorganic CRT Systems 222
8.3 Regulation and Relevant Issues 233
8.4 Summary 241
References 242
Further reading 242

Index *243*

ABOUT THE AUTHOR

Dr Xiang Zhang, the Royal Society Industry Fellow at University of Cambridge, is a materials scientist and one of the leading biomaterials and medical devices experts in the world, with 33 years combined experience; 17 years in academia and 16 years in industry. He places particular emphasis on carrying out fundamental but applied research as he believes that this in-depth fundamental understanding of specific scientific issues is the key to the design and development of successful medical products for industry to benefit society. He undertook his PhD and postdoctoral research at Cranfield University where he studied materials physics and nanofracture mechanics of organic and inorganic hybrid materials and developed new materials for ICI, the largest chemical manufacturer in Britain at that time. He was awarded an industrial fellowship at the University of Cambridge in 1995. His industry experience was gained at Abbott in 1999, where, as Principal Scientist, his work covered almost all aspects of medical materials and devices from R&D and Manufacturing. Further industrial experiences were gained with Cambridge NanoTech and Lucideon as Consultant Director and Principal Consultant, respectively, working for clients throughout the world to provide materials solutions for pharmaceutical companies and research institutes.

Dr Mark Cresswell is a Senior Scientist at Lucideon and has been instrumental in developing Lucideon's iCRT (inorganic Controlled Release Technologies) research programme. His research covers a wide spectrum of the chemical sciences which has enabled him to become an expert in his field. In addition to the inorganic chemistry discussed in this iCRT book, he has good experience in organic synthetic chemistry. This includes the development of new synthetic routes for solution-phase-based peptide synthesis and the use of a diastereoselective Pictet-Spengler reaction to generate optically pure tetrahydro-β-carbolines. Dr Cresswell later applied this methodology to the asymmetric synthesis of medicinally useful indole alkaloids. He was awarded an MChem in Chemistry with Medicinal Chemistry from the University of Manchester and he received his PhD from Keele University. At Lucideon he is responsible for the development of innovative new materials for healthcare applications with a focus on the application of novel inorganic-based drug delivery systems. Particular research interests are the use of novel sol-gel processing to prepare silicate and phosphate derived porous drug delivery systems and the design of new glasses with controlled dissolution properties.

PREFACE

The use of inorganic materials for pharmaceutical formulations is not a new concept nor is the idea of using materials such as organic-based polymers to develop controlled release formulations with beneficial properties over traditional formulations. This book's intention is to marry these two separate subjects by introducing the *new concept* of inorganic controlled release technologies, to provide readers with a theoretical and practical guide to the use and applications of this new technology. The breadth of different materials being used and the control over their properties to achieve controlled drug release is at times staggering.

By shining a spotlight on these new technologies and approaches to formulations science, it is our hope that, at the very least, the most promising materials may in time become part of the pharmaceutical laboratory's go-to materials for developing new formulations, within both industrial and academic research groups. Lucideon, amongst other research groups, is at the forefront of new materials research in this field and the responsibility is on us to raise awareness of these materials to achieve this end.

This book is divided into two main parts. The first part describes some fundamental concepts and theories of materials chemistry, physics, processing and characterization methods directly relevant for inorganic CRT materials (Chapters 1-3), which are vital to understanding the interactions between drug molecules and inorganic excipient systems and to the future practical applications of these new drug formulations. Following this is an introduction to the design and development of a range of different inorganic CRT systems (Chapters 4-7), which covers some of the excellent work being carried out in research labs globally. In this respect, we hope that this book will function as a useful reference text for these fundamental materials science concepts for professionals working for pharmaceutical companies, universities and research institutes, as well as for students who are studying in the relevant fields.

ACKNOWLEDGMENTS

We would like to thank everyone at Lucideon but in particular some members of the Healthcare team: Phil Jackson, Ian Campbell, Ben McCarthy, Gemma Budd and Matthew Cantelo, whose diligent approach and innovation in new materials research has led the development of iCRT materials. Much of the knowledge and inspiration for this book was generated by being around these highly skilled, inspirational and passionate scientists. We would especially like to thank Tony Kinsella, CEO of Lucideon, for providing us with the vision to develop Lucideon's Healthcare Division.

Special thanks must go to Professor Aldo R. Boccaccini and his team members Jasmin Hum, Elena Boccardi and Anahí Philippart from the Institute of Biomaterials, University of Erlangen-Nuremberg, who contributed the material for Chapter 5. Their expert knowledge in the theoretical basis and application of controlled release mesoporous bioactive glasses is a worthy addition to this text.

Thanks also go to the Royal Society who provides a fellowship that allows Dr Zhang to carry out fundamental research and book writing at the University of Cambridge.

A final word of appreciation must go to Xiang's wife Yong Pan and Mark's wife Gemma, whose never-ending support allowed us the opportunity to work in the least sociable hours to accomplish this book, which would have otherwise been an impossible task.

Dr Xiang Zhang and Dr Mark Cresswell,
Lucideon Cambridge Healthcare Division
April 2015

KEY FEATURES

- This will be the 1st book on inorganic controlled release technologies covering key aspects from chemistry, physics, synthetic methods, formulation design, characterization and evaluation. Case studies from an application perspective will also be included to provide practical guidance on how to use inorganic CRT as an alternative to organic polymer systems for both future drug developments and other active ingredient applications.
- Inorganic CRT offers solutions for unmet business needs for improving controlled release of actives as a promising alternative to organic (polymer) CRT. Traditional CRT systems are known to have difficulty with the controlled delivery of both poorly and highly water-soluble drug compounds, whereas inorganic CRT can deliver these two extremes of actives with a much higher level of control.

CHAPTER 1

Materials for Inorganic Controlled Release Technology

Contents

1.1 Introduction 1
1.2 Comparison Between Organic and Inorganic CRT 2
1.3 Materials Chemistry and Processing Technology 7
1.3.1 Fusion-based approach to making water-soluble glasses 7
1.3.2 Sol-gel approach 9
1.3.3 Surfactant template approach for mesoporous silica 10
1.4 Materials Physics and Drug-Loaded Micro/Nanostructure 12
References 15

1.1 INTRODUCTION

The concept of controlled release, as opposed to instant release, refers to a gradual liberation of an agent in order to maintain its concentration or availability over a period of time. This is schematically illustrated in Figure 1.1. Since the first polymer was studied for drug controlled release in the 1970s,[1] polymers have been the predominant materials used for drug controlled release applications.[2–5] This is because, when a pharmaceutical agent is encapsulated within, or attached to, a polymer, drug safety and efficacy can be improved because the active agent is released slower than with a classical drug carrier such as starch. Further development of controlled release technology (CRT) focuses on biodegradable polymers, which have clear advantages over non-biodegradable polymers that require retrieval or further manipulation after introduction into the body.

We can generally classify CRT into two basic types.[6] The first type is slow or sustained controlled release, that is, delivering an active agent over an extended period of time. The second type is triggered release, that is, controlled release of an active agent in response to environmental stimuli. Combination of the two types of CRT can generate other, more complex, CRT

Inorganic Controlled Release Technology
http://dx.doi.org/10.1016/B978-0-08-099991-3.00001-6

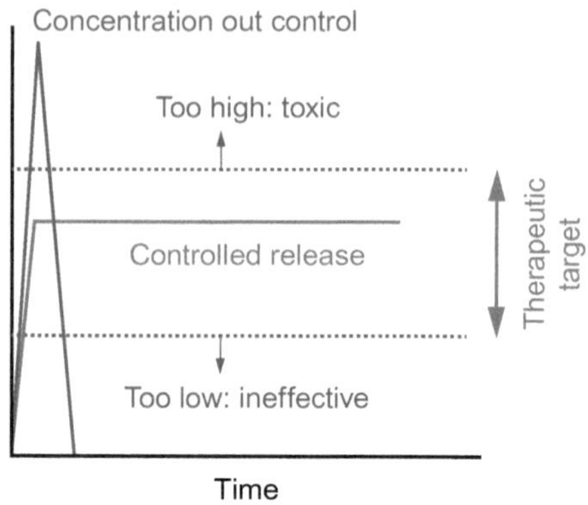

Figure 1.1 Schematic illustration of controlled and non-controlled drug release profiles.

systems. There are multiple advantages associated with a prolonged delivery of active ingredients. Firstly, a more efficient action of the agent delivered is achieved since the dosage is kept at a certain level for a continued period of time. With reference to Figure 1.1, maintaining the concentration of active ingredients within the therapeutic target range will eliminate any associated side effects, due to a too high (toxic) or too low (ineffective) dose. Additionally, keeping a steady active concentration in the bloodstream reduces the frequency of dosage, something which has a positive effect on the patient in terms of convenience and in improving their quality of life. Furthermore, as a result of the more efficient delivery and bioavailability of the active ingredients, the overall level of active ingredients employed may be reduced to achieve the same effect. In the food industry, this type of controlled release avoids or reduces the loss of ingredients, such as vitamins and minerals, during processing. Finally, it permits the separation of incompatible or unstable compounds that are only liberated when their action is required.

1.2 COMPARISON BETWEEN ORGANIC AND INORGANIC CRT

Polymeric drug delivery has been extensively studied in the past 40 years, leading to the publication of a number of reviews relevant to the commercialization of polymeric drug delivery products.[7–9] For more comprehensive reviews of the underlying science, the reader can refer to the many review articles.[10–12] The main advantage of polymeric organic CRTs over more conventional formulations is that they have allowed the development of robust sustained drug delivery products to occur. The encapsulated drug is released over extended periods of time and hence eliminates the need for multiple doses. This feature can improve patient compliance, especially for drugs for chronic indications requiring frequent injections (such as in the treatment of deficiencies of certain proteins). Polymers can also act as a shield to protect the drug from the physiological environment and hence improve its stability *in vivo*. This particular feature makes this technology attractive for the delivery of labile drugs, such as acid-sensitive small molecules and biomolecules.

One of the fundamental problems associated with organic CRT, in general, is the so-called burst effect.[13–15] That is, the drug is suddenly released in a much larger quantity than is required therapeutically without any measure of control. The mechanism of the burst effect involves solvent (such as water) swelling in the polymer phase to generate a gradual build up of internal pressure. Eventually, this pressure becomes so great that the 'wall' (or the polymeric 'skin' of the drug encapsulating formulation) suddenly breaks, resulting in the release of a larger quantity of drugs than is required. This sudden release of drugs at high concentration is potentially a cause of toxicity.[16]

A particular feature of polymeric controlled release systems is that it is very difficult to achieve good controlled release for two particular types of drugs. One of these types is very water-soluble drugs and the other is poorly water-soluble drugs. In addition to the above, the mechanisms controlling drug release from polymeric systems are often very complicated. Developing novel polymeric systems to deliver particular drug compounds is therefore a difficult and time-intensive process. These complex release mechanisms from polymeric systems may include one or a combination of bulk diffusion, surface diffusion, bulk erosion and surface erosion. In addition, the many physical variables that can influence polymer chain degradation must also be considered. This makes the design of polymeric controlled release systems and the prediction of release rates exceptionally difficult. Therefore, there are many variations and variables to consider when trying to achieve ideal control of drug release from polymeric organic CRT. These variations and variables include

- chemical composition;
- crystallinity (for crystalline polymers);
- glass transition temperature (for both crystalline and amorphous polymers);
- molecular weight;
- interaction with drugs;
- effect of size and shape of polymer matrix and drugs;

- effect of pH;
- effect of enzymes (biopolymers): there are conflicting results published on the effect of enzymes on polymer degradation mechanisms (hydrolytic vs. enzymatic cleavage), which makes the mechanisms of organic CRT even more complicated and unpredictable;
- drug load: difficult to achieve molecular dispersion but drug agglomerates/lumps of drugs are typically found within the polymer matrix; and
- toxicology: a general concern for many polymers used for CRT.

Development of new polymeric materials for controlled release applications was initially limited due to the above complexities. More recently though the number of materials used in commercial formulations has increased. Examples of these commercial products include Lupron Depot®, Zoladex®, Norplant®, Gliadel®, Port®, Oros®, EnSotrol™, Duros™, DermaSal™ and Lipoparticle™.

Inorganic materials have a longer history of being used by human beings than organic polymers. They are part of our daily dietary intake and our life is sustained by the complex interactions of inorganic and organic substances. However, studying inorganic materials for controlled release has a short history. In terms of the development of commercial products, Lucideon is a pioneer in the use of inorganic materials for controlled release applications. Lucideon's inorganic controlled release technology (iCRT) is based on over 70 years of experience in glass and ceramic materials technology. For many years now Lucideon has been producing inorganic CRT materials for commercial use, the water-soluble nature of these products ensures the controlled sustained release of key ions. This inorganic CRT has been further developed and more inorganic CRT systems have been designed and produced with tailored controlled release profiles. The release rate can be modified to achieve sustainable active release with time frames ranging from minutes, hours, days, months, or even up to years. These materials have great potential for the controlled release of drugs, offering an alternative to polymer routes or opening new doors for certain drug applications where polymeric organic CRT fails to work. To make the differences between organic polymer and inorganic CRT clear, Table 1.1 lists several key characteristics comparing the two materials as drug carriers.

Although it is perhaps worth discussing several advantages of inorganic CRT over organic CRT, only three of the more important ones are highlighted and discussed in detail here.

Simple controlled release mechanisms: One of the unique properties of inorganic CRT is that very simple release mechanisms control the rate of drug release. It is either simply (i) through dissolving of an inorganic CRT

Table 1.1 A comparison of organic and inorganic CRT formulations

Characteristic	Polymeric organic CRT	Inorganic CRT
Mechanisms	Many variables, results can be unpredictable	Simple systems primarily through carrier dissolution or active diffusion. More complex formulations are combinations of both
Processing	Both polymers and drugs are organic solvent dependent in most circumstances	
Often aqueous based, many are inert to organic solvent	Following standard processing	
Drug in the delivery system	Drugs form agglomerates, normally at, and above, micrometre scale	Drug dispersed to smaller scale, on the scale of a few nanometres to tens of nanometres
Effect of pH and temperature	Most are pH and temperature dependent for degradable polymers	Can be designed to be pH dependent
'Burst' effect	Swelling, change of structure, potentially leading to the 'burst' effect	Is minimized by homogenous drug distribution and controlling porosity
Toxicity	Potential toxicity: relict degradation products after use due to polymer chain scission	Low toxicity, biocompatible and/or bioresorbable
Solubility	Solubility variation affected by status of physical and chemical degradation	Solubility constant for a given system

drug carrier to release a drug/s in a controlled manner (such as amorphous glass material) or (ii) through drug molecular diffusion out of a porous structure (such as a porous ceramic). It is also possible to combine the two. This dissolution and/or diffusion mechanism is a simple physical process and the rate of dissolving and/or diffusion can be designed through materials formulations. This is a very important factor when considering development of inorganic CRT for a specific drug controlled release. If the drug release rate can be pre-determined, at least in theory, through formulation of drug carriers, it can greatly reduce R&D time. It is well known that the development of polymeric organic CRT systems for controlled drug release can be a lengthy process. It is anticipated that, due to their relative simplicity, inorganic CRT systems can be designed and developed over much shorter time frames. It is also worth bearing in mind, a polymer system involves both physical and chemical changes on the surface and within the polymer bulk. In addition, there are many variables and unpredictable factors that will change the course of polymer physical and chemical degradation, hence changing the drug release mechanisms and the rate/speed too, which is in contrast to the simple dissolving and/or diffusion processes controlling drug release from inorganic CRT systems.

Towards molecular levels of drug dispersion: Another unique feature of inorganic CRT is its capacity to load and disperse drug molecules on a scale of a few nanometres. In contrast, in polymeric organic CRT systems, drug molecules are normally in the form of micron-sized agglomerates. For example, inorganic CRT systems made through sol-gel technology typically have pore sizes less than 10 nm in diameter. It is almost intuitive to understand that one nano-pore space can only accommodate a small number of drug molecules and in some cases perhaps even only one. This is why the level of drug dispersion can approach the molecular level, which in turn can greatly enhance drug release efficiency. Both drug bioavailability and solubility achieved by inorganic CRT are increased dramatically in comparison with polymer CRT systems. This is because the drug loaded into the inorganic CRT system will have true molecular interaction—drug/carrier or drug/solvent (such as water) interactions. This is very different to the polymer CRT system where drug molecules are commonly dispersed to micrometre scales in the form of larger drug aggregates.

Good solutions for very water-soluble and poorly water-soluble drugs: In view of the biopharmaceutical classification system,[17] seen in Figure 1.2, inorganic CRT will provide not only alternatives to current polymer CRT systems, but also perhaps superior formulation solutions if drug solubility is a key factor to be considered. Polymer CRT systems have failed to provide effective delivery solutions for drug compounds at the two very extremes of the solubility

Class I High solubility High permeability	Class II Low solubility High permeability
Class III High solubility Low permeability	Class IV Low solubility Low permeability

Figure 1.2 Classification of drug compounds based on their *in vivo* solubility and permeability.

scale. In the case of very soluble compounds, the lipophilic nature of most polymers leads to incompatibilities between the controlled release polymer and the drug at the molecular level. On the other hand, for a poorly soluble drug, polymer CRT cannot disperse the drug in a pseudo- molecular status in the way that inorganic CRT can. In polymer CRT systems where the drug is dispersed in the formulation in micron-sized aggregates, it becomes very difficult to achieve a controlled release as the dissolution rate of these compounds precludes the available drug concentration from reaching therapeutic levels.

1.3 MATERIALS CHEMISTRY AND PROCESSING TECHNOLOGY

Inorganic drug carriers can be designed and manufactured using similar classes of ingredients as to the production of glass and/or ceramic materials. Two main possibilities exist for the production of glass and ceramic materials for controlled release applications: fusion and sol-gel routes.

1.3.1 Fusion-Based Approach to Making Water-Soluble Glasses

Traditional fusion-derived glasses are prepared by melting mixtures of the glass components in the appropriate ratios at high temperatures, followed by rapid cooling (or quenching) into a glassy solid structure. Further processing may involve grinding to obtain particles of the desired size, which can then be further sintered or shaped into the appropriate state for delivery. This synthesis route permits the incorporation of specific elements into the glass structures as active ingredients; this is in addition to their potential as matrices or carriers for other active compounds.

The basic ingredients for glass formation and their function are shown in Table 1.2. They are (i) network former, (ii) intermediate and (iii) network modifier. The fusion technology employs very high temperatures, normally above 1000 °C. The high-temperature processing of individual oxides to create a molten glass seems at first glance to be incompatible with temperature-sensitive drugs as no drugs can survive at such high temperature. However, specific compositions of glasses can be prepared in such a way that following their production, they can be treated at certain temperatures to induce distinct regions of phase separation within the bulk material. Careful choice of the necessary conditions then allows selective removal of the least chemically durable phase, which then leaves behind a porous structure. This porous structure is ideally suited for the adsorption of active drug molecules to take place. This elegant solution to producing porous glasses is demonstrated in Figure 1.3.

Table 1.2 Fusion-based glasses: the basic constituents

Structural role	Oxide	Function
Network former	SiO_2	Enhances 'polymeric' network to reduce solubility
	B_2O_3	
	P_2O_5	Modifies solubility & pH, major bone constituent
Intermediate	Al_2O_3	
	TiO_2	
	ZrO_2	
Network modifier	MgO	Modest network disruption; micronutrient
	CaO	Modest network disruption; alkaline bone constituent
	SrO	Synergistic effect with CaO
	BaO	
	PbO	
	ZnO	Synergistic effect with CaO
	Na_2O	Disrupts network to increase solubility & pH
	Li_2O	As above
	K_2O	As above

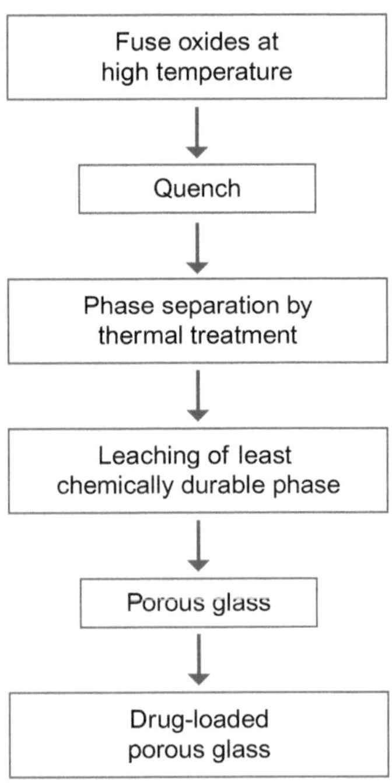

Figure 1.3 Production of drug-loaded porous glasses.

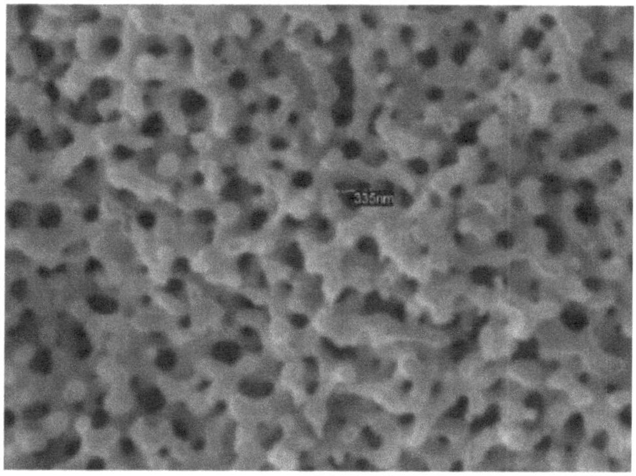

Figure 1.4 SEM of a phase separated glass. *(Courtesy of Lucideon).*

1.3.2 Sol-Gel Approach

Sol-gel techniques comprise the transformation from a colloidal suspension containing the glass precursors into a solid network. By adding sacrificial agents it is possible to obtain porous structures that can be loaded with the active agent to be released. In comparison with fusion-derived glasses, sol-gel glasses and ceramics display bioactive properties that are not solely determined by the composition and the choice of the precursor materials, but also by the final porous and ordered structure.

Distinct from the fusion-based approach, the sol-gel synthesis route is performed at low temperatures, typically at or around room temperature. The basic chemical reactions for a simple pure silica sol-gel are shown in Figure 1.5. Depending on whether acid or base is used to catalyze the reaction (and indeed, depending on many other factors such as the catalyst molarity), the preference for hydrolysis versus condensation reactions can be altered. This in turn affects the final pore structure.

Figure 1.6 illustrates the different ways in which the sol-gel reaction can be manipulated to produce products with different structures, morphologies and physical properties. It is possible to introduce drug compounds into the porous sol-gel products *via* an adsorption process. Loading the sol-gel materials with drug compounds in this way enables the incorporation of active molecules with a range of physicochemical properties. The specific interaction energies between the drug, the adsorption solvent and the sol-gel

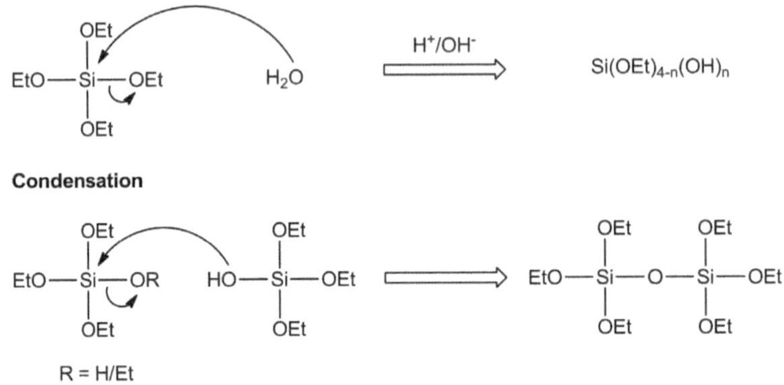

Figure 1.5 Formation of silica sol-gel from tetraethyl orthosilicate (TEOS).

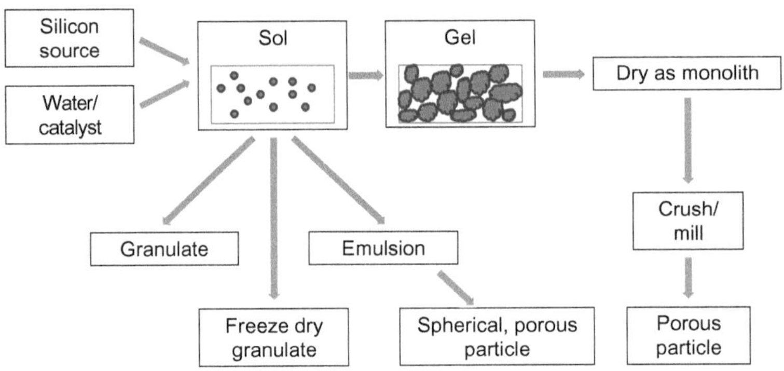

Figure 1.6 Processing of silica sol-gels.

surface are critical in determining the extent to which the molecules are deposited within the porous structure or adsorbed to the particle surfaces.

By employing other precursors, apart from silica, in the form of alkoxides or metal salts, multi-element oxide sol-gels can be formed to introduce and allow variation in properties such as bioactivity and solubility. Examples of the basic constituents are listed in Table 1.3.

1.3.3 Surfactant Template Approach for Mesoporous Silica

To date, mesoporous silicas are by far the most well-studied inorganic materials for potential use as controlled release drug delivery systems. Some detailed science and technological applications of mesoporous silica will be introduced in Chapter 4.

Table 1.3 Sol-gel technology: the basic constituents

Structural role	Oxide	Additional functionality
Network former	SiO_2	Enhances 'polymeric' network to reduce solubility
	B_2O_3	
	P_2O_5	Modifies solubility & pH; major bone constituent
Intermediate	TiO_2	
	ZrO_2	Provides radio-opacity
Network modifier	MgO	Modest network disruption; micronutrient
	CaO	Modest network disruption; alkaline bone constituent
	SrO	Synergistic effect with CaO; hard tissue healing
	ZnO	Synergistic effect with CaO; hard tissue healing
	Na_2O	Disrupts network to increase solubility & pH
	K_2O	Disrupts network to increase solubility & pH
	Fe_2O_3	Preparation of magnetic nanoparticles for site-specific targeting

Figure 1.7 shows a schematic synthesis route of mesoporous silicates through a surfactant template approach. Mesoporous silica is normally synthesized in this method by the polycondensation of silicate species, originating from different sources of silica (e.g., sodium silicate, tetraethyl orthosilicate (TEOS)).[18] Synthesis conditions such as source of silica, type of surfactant, ionic strength, pH and composition of the reaction mixture, temperature and duration of synthesis affect the surfactant micellar conformation, the silica-surfactant interactions and the degree of silica polycondensation. These conditions determine the characteristics of the porous

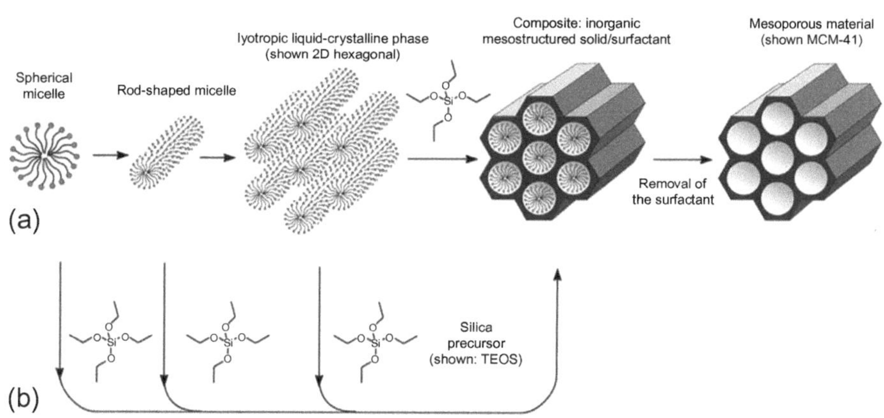

Figure 1.7 Schematic synthesis route of mesoporous silica using a surfactant template. *(Reprinted with permission from Ref. 21. Copyright 1992, American Chemical Society).*

structure (type of mesostructure, diameter and volume of the pores, wall thickness) and the macroscopic morphology.[19,20]

A wide variety of ionic and non-ionic surfactants have been used for obtaining materials with different porous and morphological characteristics. MCM-41, MCM-48 and MCM-50 mesoporous silicas (MCM: Mobil Composition of Matter) with hexagonal, cubic and lamellar mesostructures, respectively, have been synthesized by using alkyl ammonium surfactants and TEOS or sodium silicate in a basic medium.[21] Their pore size and wall thickness do not go beyond 4.0 and 2.0 nm, respectively. By using cationic surfactants in acidic medium the first SBA-type materials with different mesostructures were obtained, with porous characteristics similar to the MCM-X-type materials.[22] SBA-15 and SBA-16 silica (SBA: Santa Barbara University) with larger pore sizes and thicker walls are prepared by using non-ionic surfactants derived from poly(propylene oxide) and poly(ethylene oxide) in acidic medium.[23]

1.4 MATERIALS PHYSICS AND DRUG-LOADED MICRO/NANOSTRUCTURE

For drug controlled release, glass and ceramic materials have unique properties that make them promising candidates as drug carriers.[6] In general, they are thermally and chemically resistant, especially in strong acid and alkaline environments, which will help prevent degradation of drug active ingredients. They do not react with solvents (e.g., gastric resistance) to which most polymers are susceptible. This opens up, for instance, new delivery routes, such as targeted enteral delivery. Also, compared to polymers, they do not swell or change structure under pH or temperature variations, avoiding potential dose-dumping problems. Upon processing into a microporous or nanoporous form, they provide numerous sites for hosting functional compounds. Depending on their composition, the solubility of glasses and ceramics can be tailored to suit the requirements of controlled delivery. Figures 1.8 and 1.9 show models and representative microscopy images for drug-loaded glasses and ceramics.

Since drug-loaded particles appear particularly suited for controlled release and specific site targeted release, they have been a major focus of research for new drug delivery systems.[24] Among these drug delivery systems, inorganic porous materials are emerging as a new category of host/guest systems. MCM-41 and SBA-15 are typical examples of mesoporous templated silica that have been widely investigated. They are comprised

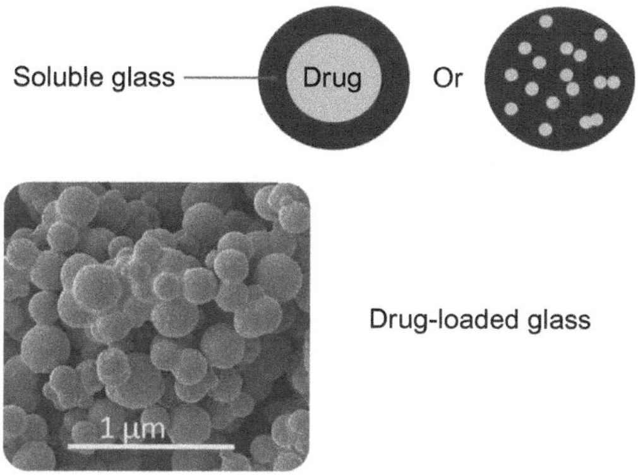

Figure 1.8 Model and microscopy image of a drug-loaded glass.

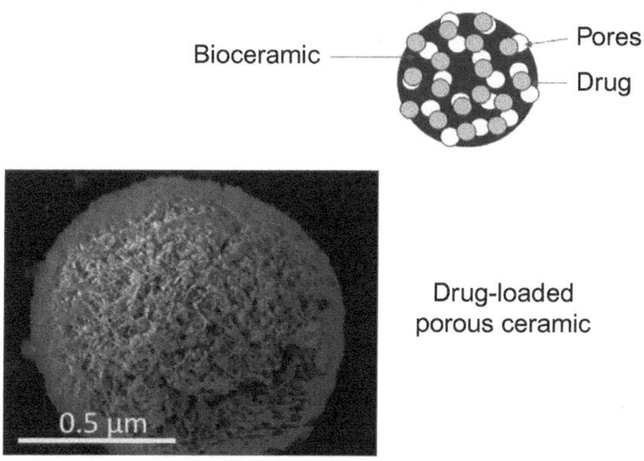

Figure 1.9 Model and microscopy image of a drug-loaded porous ceramic.

of a honeycomb-like porous structure containing empty channels (mesopores) that are able to absorb/encapsulate relatively large amounts of bioactive molecules. The unique properties include high surface area (>900 $m^2\ g^{-1}$), large pore volume (>0.9 $cm^3\ g^{-1}$) and tunable pore size with a narrow distribution (2-10 nm). These features are shown in Figure 1.10 with a model of mesoporous silica loaded with a drug. Silica in general has good chemical and thermal stability, which makes it potentially suitable for various controlled release applications. Highly ordered, mesoporous silicas appear very suitable as host materials and important applications are being developed in areas such as heterogeneous catalysis, adsorption, ion exchange and inclusion chemistry. On account of their well-defined and large inner porosity, and consequently their large surface area, templated mesoporous silicas exhibit structural properties required for

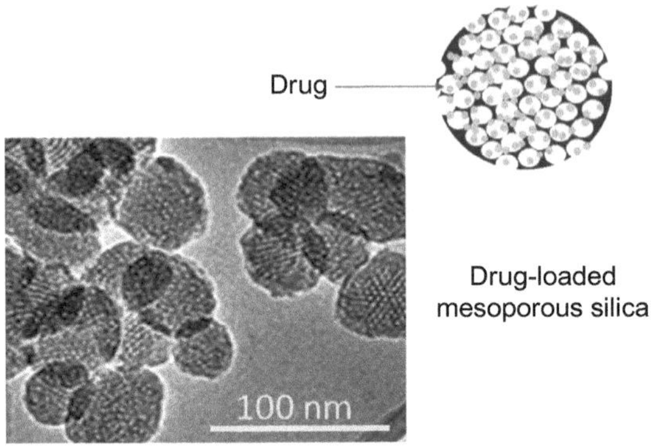

Figure 1.10 Model of drug-loaded mesoporous silica and a typical TEM image. (*Microscopy image reproduced courtesy of Prof. Aldo Boccaccini, University of Erlangen-Nuremburg*).

drug carriers. Moreover, silica, such as Aerosil®, is a widely used pharmaceutical excipient; therefore, it would be interesting to use ordered mesoporous silicas in alternative formulations for drug delivery.

Sol-gel-derived mesoporous silica nanoparticles synthesized under mild conditions are also of interest due to the relatively simple approach to production, the ease with which chemical and physical modification can take place, and the capacity to maintain the functional activity of bioactive agents.[25] The unique mesoporous structure of silica facilitates effective loading of drugs and their subsequent controlled release. The properties of the mesopores, including pore size and volume as well as the surface properties, can be altered depending on the additives used to fabricate the mesoporous silica nanoparticles. The presence of a chemically active surface enables functionalization to take place with a view to modify the surface properties and also to form interactions between the active molecules and the silica surface. The tunable mesoporous structure and modifiable surface of mesoporous silica nanoparticles allow incorporation of various classes of drug molecules and controlled delivery to the target sites and have a great potential for a variety of drug delivery applications, such as the site-specific delivery and intracellular controlled release of drugs, genes and other therapeutic agents.[26]

Nanosized mesoporous silica particles with high colloidal stability are attracting growing attention as drug delivery systems for targeted cancer treatment and as bioimaging devices.[27] These new developments demonstrate the high potential of nanosized mesoporous silicas as they have high drug loading capacity and multiple functionality for targeting and entering different types of cells. With the aid of an efficient stimuli-responsive release, mesoporous silicas can effectively be used to release chemotherapeutic compounds within or near cancer cells. Mesoporous silica nanoparticles are viewed as a promising and flexible platform for numerous biomedical applications.

REFERENCES

1. Langer R, Folkman J. Polymers for the sustained release of proteins and other macromolecules. *Nature* 1976;**263**:797–800.
2. Siepmann J, Elkharraz K, Siepmann F, Klose D. How autocatalysis accelerates drug release from PLGA-based microparticles: a quantitative treatment. *Biomacromolecules* 2005;**6**(4):2312–9.
3. Ahmad Z, Khuller GK. Alginate-based sustained release drug delivery systems for tuberculosis. *Expert Opin Drug Deliv* 2008;**5**(12):1323–34.
4. Klose D, Siepmann F, Elkharraz K, Siepmann J. PLGA-based drug delivery systems: importance of the type of drug and device geometry. *Int J Pharm* 2008;**354**(1-2):95–103.
5. Batycky RP, Hanes J, Langer R, Edwards DA. A theoretical model of erosion and macromolecular drug release from biodegrading microspheres. *J Pharm Sci* 1997;**86**(12):1464–77.
6. Campbell I. Lucideon Ltd. Towards the future of controlled release applications, www.lucideon.com [accessed 02.04.15].
7. Gombotz WR, Pettit DK. Biodegradable polymers for protein and peptide drug delivery. *Bioconjugate Chem* 1995;**6**(4):332–51.
8. Sinha VR, Khosla L. Bioabsorbable polymers for implantable therapeutic systems. *Drug Dev Ind Pharm* 1998;**24**(12):1129–38.
9. Langer R. Drug delivery and targeting. *Nature* 1998;**392**(6679 Suppl.):5–10.
10. Tanguay JF, Zidar JP, Phillips III HR, Stack RS. Current status of biodegradable stents. *Cardiol Clin* 1994;**12**(4):699–713.
11. O'Donnell PB, McGinity JW. Preparation of microspheres by the solvent evaporation technique. *Adv Drug Deliver Rev* 1997;**28**(1):25–42.
12. James HP, John R, Alex A, Anoop KR. Smart polymers for the controlled delivery of drugs—a concise overview. *Acta Pharm Sin B* 2014;**4**(2):120–7.
13. Allison SD. Analysis of initial burst in PLGA microparticles. *Expert Opin Drug Deliv* 2008;**5**(6):615–28.
14. Xiang Z, Sarazin P, Favis BD. Controlling burst and final drug release times from porous polylactide devices derived from co-continuous polymer blends. *Biomacromolecules* 2009;**10**(8):2053–66.
15. Huang X, Brazel CS. On the importance and mechanisms of burst release in matrix-controlled drug delivery systems. *J Control Release* 2001;**73**(2-3):121–36.
16. Jämstorp E, Forsgren J, Bredenberg S, Engqvist H, Strømme M. Mechanically strong geopolymers offer new possibilities in treatment of chronic pain. *J Control Release* 2010;**46**(3):370–7.
17. Amidon GL, Lennernäs H, Shah VP, Crison JR. A theoretical basis for a biopharmaceutic drug classification system: the correlation of *in vitro* drug dissolution and *in vivo* bioavailability. *Pharm Res* 1995;**12**(3):413–20.
18. Giraldo LF, López BL, Pérez L, Urrego S, Sierra L, Mesa M. Mesoporous silica applications. *Macromol Symp* 2007;**258**(1):129–41 [Special Issue: Polychar-15 World Forum on Advanced Materials].

19. Di Renzo F, Testa F, Chen JD, Cambon H, Galarneau A, Plee D, et al. Textural control of micelle-templated mesoporous silicates: the effects of co-surfactants and alkalinity. *Microporous Mesoporous Mater* 1999;**28**(3):437–46.
20. Tanev PT, Pinnavaia TJ. Mesoporous silica molecular sieves prepared ionic and neutral surfactant templating: a comparison of physical properties. *Chem Mater* 1996;**8**(8):2068–79.
21. Beck S, Vartuli JC, Roth WJ, Leonowicz ME, Kresge CT, Schmitt KD, et al. A new family of mesoporous molecular sieves prepared with liquid crystal templates. *J Am Chem Soc* 1992;**114**(27):10834–43.
22. Huo Q, Margolese DI, Stucky GD. Surfactant control of phases in the synthesis of mesoporous silica-based materials. *Chem Mater* 1996;**8**(5):1147–60.
23. Zhao D, Huo Q, Feng J, Chmelka BF, Stucky GD. Non-ionic triblock and star diblock copolymer and oligomeric surfactant syntheses of highly ordered, hydrothermally stable, mesoporous silica structures. *J Am Chem Soc* 1998;**120**(24):6024–36.
24. Charnay C, Bégu S, Tourné-Péteilh C, Nicole L, Lerner DA. Inclusion of ibuprofen in mesoporous templated silica: drug loading and release property. *Eur J Pharm Biopharm* 2004;**57**(3):533–40.
25. Kwon S, Singh RK, Perez RA, Abou Neel EA, Kim HW, Chrzanowski W. Silica-based mesoporous nanoparticles for controlled drug delivery. *J Tissue Eng* 2013;**4**:http://dx.doi.org/10.1177/2041731413503357.
26. Slowing II, Vivero-Escoto JL, Wu CW, Lin VS. Mesoporous silica nanoparticles as controlled release drug delivery and gene transfection carriers. *Adv Drug Deliv Rev* 2008;**60**(11):1278–88.
27. Argyo C, Weiss V, Bräuchle C, Bein T. Multifunctional mesoporous silica nanoparticles as a universal platform for drug delivery. *Chem Mater* 2014;**26**(1):435–51.

CHAPTER 2

Materials Fundamentals of Drug Controlled Release

Contents

2.1 Introduction of Materials Nanostructure 17
  2.1.1 The structure of amorphous materials 18
  2.1.2 Theories of amorphous materials 21
    2.1.2.1 Glass transition *21*
    2.1.2.2 Free volume theory *22*
2.2 API Distribution Within Inorganic Matrices 23
  2.2.1 Traditional API distribution 23
  2.2.2 API distribution within inorganic CRT matrices 26
2.3 Basic Understanding of Potential Molecular Interactions 29
  2.3.1 Classical API excipients 29
  2.3.2 Interactions between API and inorganic CRT matrix systems 30
  2.3.3 The surface chemistry of silica 32
  2.3.4 Molecular interaction with directionally templated mesoporous silica systems 33
  2.3.5 Towards molecular dispersion and distribution 38
  2.3.6 Molecular interaction sites on sol-gel silica and phosphate glass 39
  2.3.7 Dissolution of phosphate glass 45
  2.3.8 Glass formulation for inorganic CRT 47
2.4 Theory and Practical Modelling of Drug Controlled Release Kinetics 48
References 53
Further reading 55

2.1 INTRODUCTION OF MATERIALS NANOSTRUCTURE

In Section 1.2 materials chemistry and processing technology were introduced and discussed in the context of the two main production routes to make glass materials for controlled release applications, that is, fusion and sol-gel. Due to its cultural and historical significance the word 'glass' is ubiquitous in many publications and is synonymous with a certain type of glassy material

Inorganic Controlled Release Technology
http://dx.doi.org/10.1016/B978-0-08-099991-3.00002-8

based around soda-lime glass—container-ware and window glass being the most common examples. However, glassy materials made by the two different processes, fusion and sol-gel, have different structural properties as a consequence of the molecular process involved during their formation. To fully understand the fundamentals of inorganic controlled release technology (inorganic CRT) materials science, it will benefit the reader of the book to introduce the theories and fundamentals of amorphous materials, in particular, on the physical aspects.

2.1.1 The Structure of Amorphous Materials

There is a general consensus that the word 'glass' signifies only that transparent material, which is mainly made from silica plus small amounts of other oxides, and is used for producing windows. We will transform this simple understanding on transparent materials into fundamentals of amorphous materials in view of their nano/microstructure.

Amorphous materials are generally regarded as non-crystalline materials. They are not new; the iron-rich siliceous amorphous materials recovered from the moon by the Apollo mission are some billions of years old. The history of man-made amorphous materials, the glass from silica, has been known for at least several thousand years. Although the science of the materials chemistry is well established, the material physics is not so well studied and understood to date. There is growing interest in the fundamental science of amorphous materials because most recently there has been great interest for their use for novel applications and technological areas. Healthcare is one of these areas including the materials science and technology for inorganic CRT, discussed here.

In general, non-crystalline amorphous materials possess randomness to some degree. The questions are what kind of randomness is present, and what is the best way to define it? There needs to be a standard based on which we can define amorphous materials and their structure. This standard is the perfect crystal, with the following definition:

> *A perfect crystal is that in which the atom or group of atoms are arranged in a pattern that repeats periodically to infinite extent.*

In this book, we regard the repeating periodicity that is in all three dimensions. However, two-dimensional (2D)[1,2] and one-dimensional crystals[3] have been reported, in order to simplify the classical definition of a perfect crystal. The basic element of the perfect crystal is an atom or group of atoms, upon which the structure of the perfect crystal is built starting at the nanometre scale and growing periodically by an infinite extent and ending at surfaces (non-infinite extent), or at a defect, that is, foreign atoms or a dislocation of the crystal.

Considering the standard definition of a perfect crystal, we now have the following definition for an amorphous structure:

> *An amorphous structure does not possess the long-range periodicity characteristics of a perfect crystal in that atoms or a group of atoms are arranged in a random pattern that cannot repeat periodically.*

To understand the nature of long-range periodicity, Figure 2.1 shows schematically the periodicity of a perfect 2D crystal and the randomness of a 2D amorphous structure. It is seen that the non-crystalline structure consists of no translational periodicity (Figure 2.1b), which is in contrast to the crystalline structure where there is such translational periodicity (Figure 2.1a). Understanding the structural arrangement of atoms in a solid substance is essential for the understanding of its physical properties, and this is true for both crystalline and amorphous materials. Determination of a crystalline structure is made straightforward by measuring a 'unit cell', containing relatively few atoms in most cases. This unit cell is the fundamental building block upon which crystals are formed: by repeating in a periodical fashion the position of the unit cell in space. Such a procedure is impossible for a non-periodic amorphous solid, for which the unit cell is

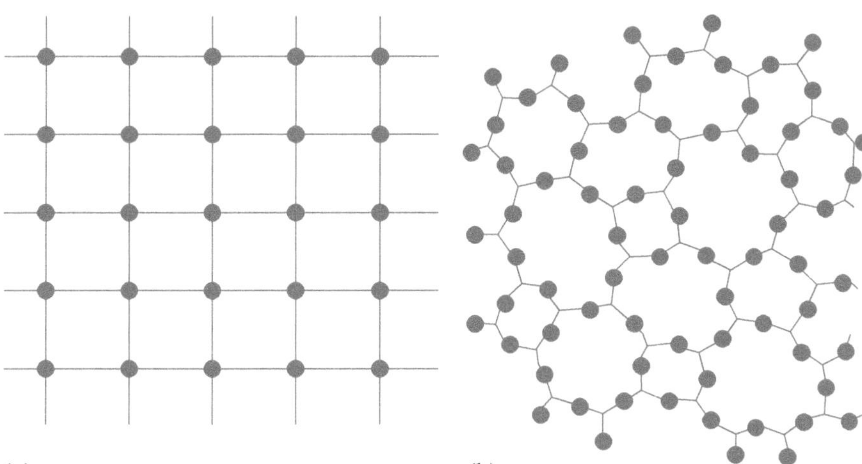

(a) (b)

Figure 2.1 Schematic representation of a 2D crystal (a) and a 2D amorphous structure (b).

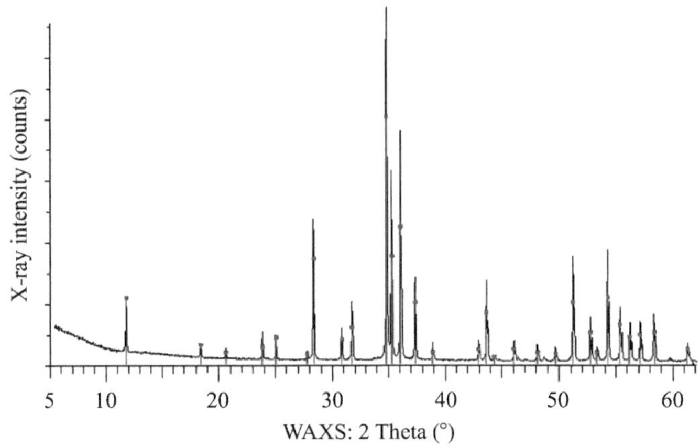

Figure 2.2 Typical crystal WAXS profile: hydroxyapatite. *(Courtesy of Lucideon X-ray Department).*

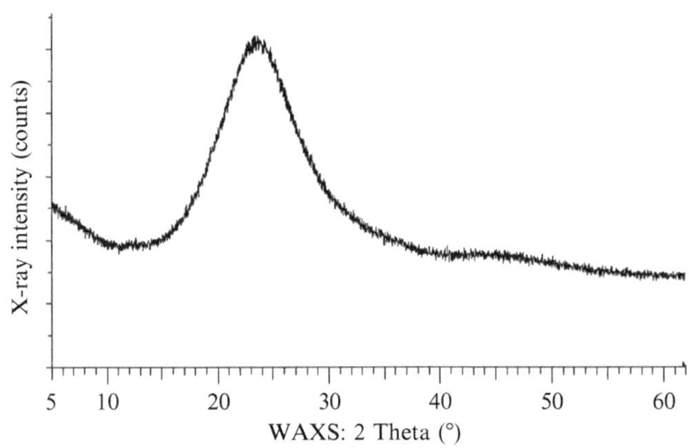

Figure 2.3 Typical amorphous WAXS profile: sol-gel glass. *(Courtesy of Lucideon X-ray Department).*

regarded as being infinitely large. Figure 2.2 shows a typical crystalline amorphous X-ray diffraction (XRD) profile, and Figure 2.3 shows an amorphous material XRD spectrum (detailed analysis of XRD will be introduced in Chapter 3).

The relative simplicity of the amorphous material XRD diffractogram belies the actual complexity of the system. The forms of randomness with which we are concerned create more uncertainty than a perfect crystal. Imaging the huge number of forms, scales and extents of amorphous randomness are all variables in view of the nanostructure. The structural variation of an amorphous material will at least affect the location, distribution, and interaction of any active ingredients present within the delivery matrices used in inorganic CRT applications.

In reality, the structure of amorphous materials is not complete randomness; they often have short-range order. Further still, some amorphous materials have considerable short-range order and others have very little. However, both have no long-range order and periodicity; for example, the structure of silica and silicates in the glassy state.

Following detailed analysis of amorphous structures, we now say that the word glass no longer means transparent window glass but amorphous materials that have no long-range atomic periodicity and a random or disordered nanostructure.

2.1.2 Theories of Amorphous Materials

Theories of amorphous materials have been an active research area with regard to physical and chemical structures and the relationship between the two. This is the key to understanding a range of physical phenomena. In this book, we will introduce two theories that are relevant to the inorganic CRT materials technology.

2.1.2.1 Glass Transition

The theory on glass transition is historically relevant to the study of glass processing,[4] but now it is applicable to all amorphous materials including polymeric materials. The glass transition is the phenomenon in which a solid amorphous phase exhibits more or less abrupt changes in derivative thermodynamic properties such as heat capacity or thermal expansion: when an amorphous material changes from a solid state to a liquid state on heating or from a liquid to a solid-like state on cooling. A glass is an amorphous solid which exhibits a glass transition. Figure 2.4 schematically illustrates the glass transition in two conditions: one is fast cooling and the other slow. It is seen that the faster cooling rate gives a higher glass transition temperature T_g. It is worth noting the differences in specific volume of the two glass products derived from the same starting material which have different thermal histories.

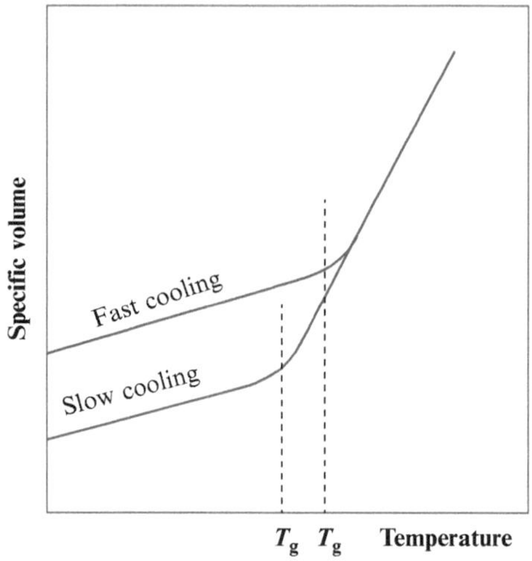

Figure 2.4 Schematic illustration of glass transitions of typical amorphous materials.

2.1.2.2 Free Volume Theory

Glass transition theory described above has demonstrated time-dependent characteristics as shown in Figure 2.4. Its importance can be seen in the differences in product specific volume, which is cooling rate dependent. This unique characteristic is applicable to all amorphous materials, and it is the glass transition that has led to the 'free volume' theory. Take an amorphous material as a 'frozen' liquid; the total volume of the liquid is divided into two parts: one is occupied by molecules or the chain of molecules, such as SiO_2, and the other part is unoccupied 'free volume'. The question is how much free volume is available? The importance being that the free volume could be used for accommodating drug molecules if we are going to use porous amorphous materials as a controlled release drug delivery system. Let us take silica SiO_2, the most investigated inorganic material for drug controlled release applications, as an example. In order to work out the free volume in a silica system, we need to be able to determine the properties of a silica standard which has zero free volume. This is almost an impossible task but we can take Stishovite as a base material for comparison purposes, which is an extremely dense tetragonal form (polymorph) of silicon dioxide of density 4.287 g cm^{-3}. For a given weight of silica, the relative free volume V_{free} can be calculated using Equation (2.1).

$$V_{free} = \frac{V - V_{st}}{V} = \frac{\rho_{st} - \rho}{\rho_{st}} \tag{2.1}$$

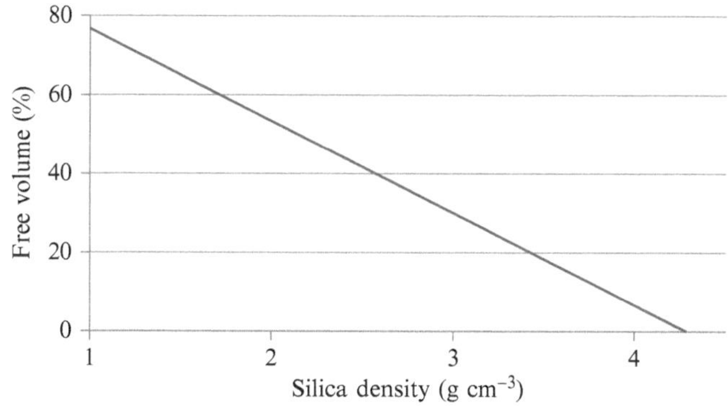

Figure 2.5 Silica free volume as a function of silica density. *(Taking Stishovite as a baseline of density 4.287 g cm^{-3}).*

where V and V_{st} are the volumes of silica and Stishovite, respectively, and ρ and ρ_{st} are the corresponding densities.

Figure 2.5 uses Equation (2.1), plotting percentage of free volume as a function of silica density. It demonstrates that most silica has large amounts of free volume that is not occupied by the SiO_2 molecules or the chain of SiO_2 molecules. As an example, let us take a porous silica made using the sol-gel technology introduced in Section 1.3.2, Chapter 1. The density of the silica is about 2.2 g cm^{-3}, which means that the silica made from the sol-gel technology has nearly 50% free volume that is not occupied by individual molecules or chains of SiO_2. This free volume is the area that has a great potential for loading active ingredients, such as drug molecules, for controlled release applications. Please bear in mind that the free volume is the total volume which combines the free volume surrounding SiO_2 molecules or the chains of SiO_2 and nanopores in the sol-gel matrix. The sol-gel technology introduced in Chapter 1 normally produces amorphous silica with pores of a few nanometres. Detailed characterization on measuring nanopores will be introduced in Chapter 3.

2.2 API DISTRIBUTION WITHIN INORGANIC MATRICES

2.2.1 Traditional API Distribution

Before discussing the features of API (active pharmaceutical ingredient)-loaded inorganic matrices, let us review the area in the context of more traditional pharmaceutical formulations using the distribution of API molecules within a tablet as an example.

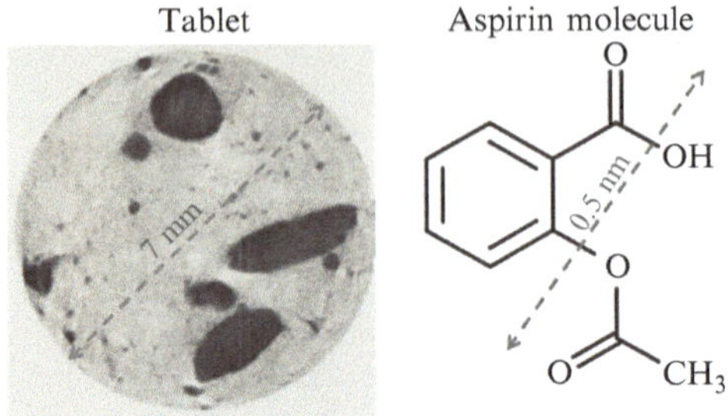

Figure 2.6 Aspirin (black phase) distribution within the tablet matrix.

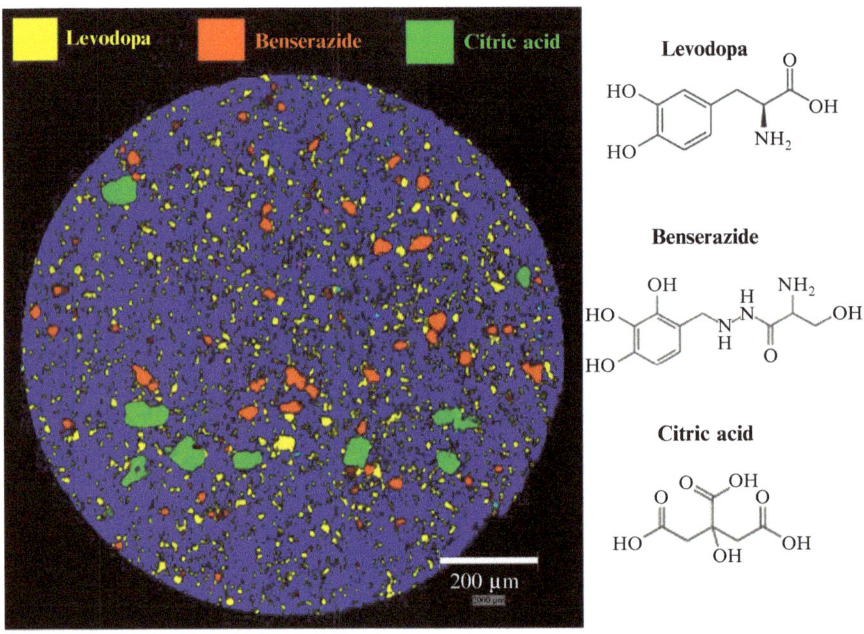

Figure 2.7 Multi API distribution—an example of a tablet used for the treatment of Parkinson's disease. *(Image taken with a Renishaw inVia Raman microscope and adapted with permission from Renishaw; http://www.renishaw.com/en/pharmaceuticals-7977).*

Figure 2.6 is a microscopy image of an aspirin tablet (left) together with the molecular structure of aspirin (right). The dark phase in the image represents aspirin crystals. It is seen that there is a range of sizes, some of which are up to a millimetre in some dimensions. Figure 2.7 is another example, showing a Raman image of a tablet used for the treatment of Parkinson's disease with multiple APIs and excipients distributed in the microcrystalline cellulose matrix. The yellow and red phases are the APIs levodopa and benserazide, respectively, and the green is citric acid, which is normally used as a disintegrant to aid drug release on contact with moisture. Ideally the citric acid should be homogeneously dispersed throughout the tablet or closely associated with the API molecules, but as can be seen in Figure 2.7, neither of these scenarios is the case. Although these representative microscopy images are not typical of all tablet formulations, it is common that drug distribution is always in the form of particles or particle agglomerates. This is not only determined by the properties of the drug molecules and excipients but more importantly, by the manufacturing technology employed.

Figure 2.8 shows schematically the general flowchart of processes employed during tablet production. It is clear that the overall tablet production is dominated by physical and mechanical processing steps, including weighing, milling, granulation, drying, blending, lubrication, compression and sometimes coating. The nature of this manufacturing technology makes it very difficult to achieve uniform API distribution with nanometre resolution, which would be ideally applicable to poorly soluble drug compounds.

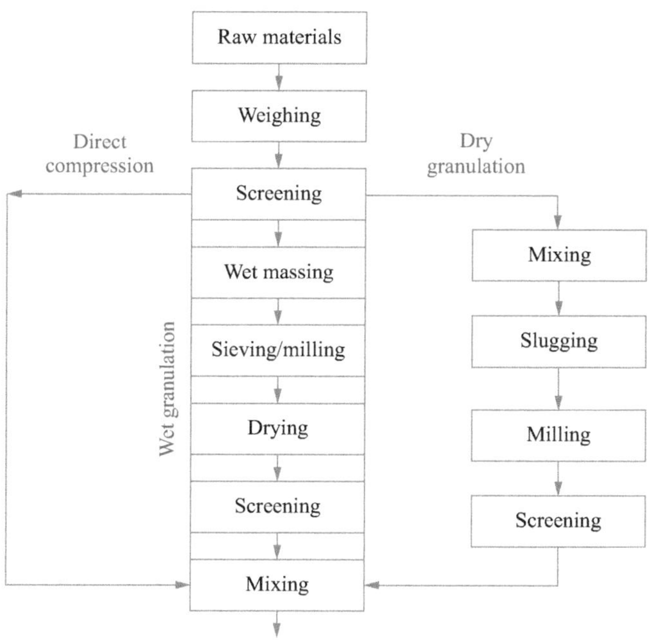

Figure 2.8 Typical tablet processing flow chart showing direct compression, wet and dry granulation. *(Adapted from Ref. 5 with permission from Moksha Publishing House).*

2.2.2 API Distribution Within inorganic CRT Matrices

One of the fundamental differences of inorganic CRT systems discussed in this book is that by using sol-gel technology, drug molecules can be incorporated within the delivery systems by two significantly different methods. The drug can be present during the nanopore forming process or alternatively the nanoporous structure can be formed first followed by the introduction of the drug molecules—such as is the case with sol-gel-derived mesoporous silicas and fusion-derived glassy materials.

To make this concept clearer it will be useful to compare the technology of classical drug manufacturing, such as that shown in Figure 2.8, with the new inorganic CRT technology introduced in Chapter 1: of the fused glass technology shown in Figures 1.3 and 1.4, the sol-gel technology shown in Figure 1.6, and the mesoporous silica technology in Figure 1.7 (and more to be introduced in Chapters 4-7 later in this book). Classical pharmaceutical manufacturing technology typically cannot produce formulations with the API well distributed within the delivery matrix with good uniformity and smaller sizes at the nanometre scale. Figures 2.6 and 2.7 are good examples where APIs are distributed not only in a non-uniform manner, but also as large agglomerates, from micrometre up to millimetre scales. Contrary to this, the new inorganic CRT is able to produce formulations with API distributed homogeneously at the nanometre level. Two of the three technologies introduced can be used to develop nanopores with good control. In this respect, the API distribution will be mainly decided by the nanostructural network, that is, the sizes and distribution of the nanopores, as this will determine where the drug molecules will reside. So in order to fully understand and study the API distribution within inorganic CRT systems, it must first be understood how to investigate the nanopore forming processes.

In the majority of instances if the inorganic CRT matrix is performing its function suitably as designed, it would be very difficult to visualize the API and its distribution within the different inorganic matrices described in this book. This is because the as-formed API phases are too small to be seen. Figure 2.9 is a scanning electron microscope (SEM) image of ibuprofen-loaded sol-gel glass particles.[6] In this case, the drug was dissolved into a silicon alkoxide solution which preceded gel formation. The author was trying to find drug phases at much higher SEM magnification but with no success. This is because the ibuprofen phase was distributed at a physical level beyond the resolution limits of a typical SEM. Figure 2.10 is another SEM image for another ibuprofen-loaded sol-gel glass. The sol-gel glass appears like a

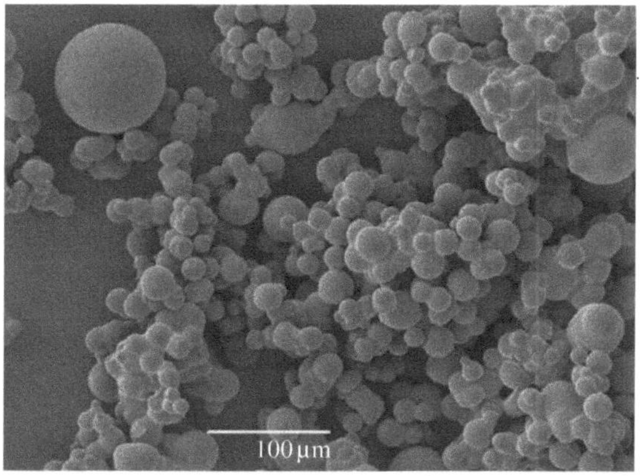

Figure 2.9 Microscopy image of ibuprofen-loaded sol-gel glass particles. *(Courtesy of Lucideon).*

Figure 2.10 Microscopy image of ibuprofen-loaded sol-gel loose particles (or micro foam). *(Courtesy of Lucideon).*

micro-foam, which is much less densified than the sol-gel particles shown in Figure 2.9. For the same reason it is impossible to see the drug ibuprofen phase even at much higher SEM magnification. The basic fact is that the drug was well dispersed into the sol-gel silica matrix at nanometre scale.

Another example is mesoporous silica. It would certainly not be possible to see API phases using the same SEM technology. For these materials, where there are controlled uniform pore size distributions, the drug molecules are located within nanopores 2-50 nm in diameter.[7-10] It is understandable that the drug phase cannot be bigger than the size of the

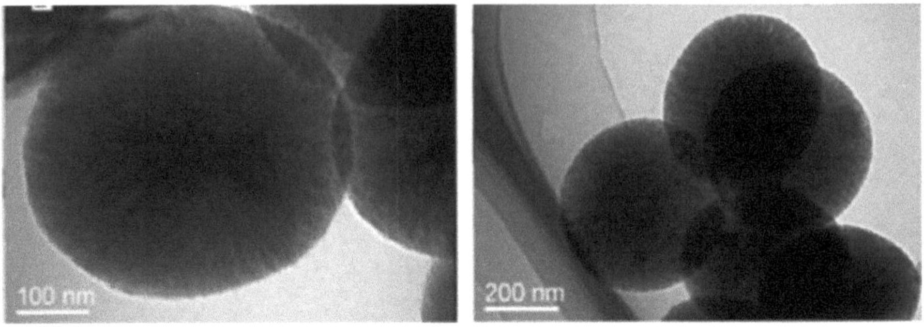

(a) (b)

Figure 2.11 TEM images of: a) mesoporous silica and b) drug-loaded mesoporous silica. *(Reprinted from Ref. 11 with permission from Elsevier).*

nanopores in mesoporous silica, which is beyond the limits of the SEM technique. Figure 2.11b is a TEM image showing a mesoporous silica carrier system loaded with a drug.[11] From the microscopy analysis no drug phases can even be seen at this much increased image resolution. This particular image at least indicates that the drug distribution is very uniform and that the size of the drug phase is around a few nanometres, in alignment with the size of the nanopores. There is good reason to expect that mesoporous silica systems will have good potential to achieve API distribution close to the molecular level.

One of the obvious advantages with reduced API particle sizes is the associated increase in the dissolution rate, and as such, the bioavailability for poorly water-soluble drugs (assuming permeability is not an issue). The dissolution rate of a drug is a function of its intrinsic solubility and its particle size. The majority of classical drug processing technologies use mechanical tools to reduce particle size such as mixing and milling (refer to Figure 2.8). This size reduction should lead to an increase in specific surface area; thus, an improvement in bioavailability could be expected. If agglomeration of particles occurs in the gastrointestinal tract then the effective surface area for dissolution will be reduced. With this in mind, mesoporous silica can dramatically increase a drug's dissolution rate and oral bioavailability. Ahern compared several loading processes in terms of drug distribution throughout the mesoporous silica matrix.[12] A poorly soluble drug was deposited into the mesoporous structure of SBA-15 with success. XRD confirmed that there is no drug crystalline structure detected, which indicates that the drug should have achieved pseudo-molecular dispersion and distribution.

The outcomes of API nano-dispersion and nano-distribution in an inorganic CRT system will certainly change the physicochemical performance of a drug and the effectiveness of any given medication. Typical particle sizes of a nano-dispersed API could be 1×10^3-1×10^6 times smaller than with classical formulations, which are often at micro- and millimetre scales. This new characteristic will provide beneficial properties to a range of new applications, such as chemotherapeutic agents for cancer treatments.

A challenge for cancer therapy is the limitation of effective biocompatible delivery systems for most hydrophobic therapeutic anticancer drugs.[13] It is particularly important to improve the aqueous solubility of poorly soluble drugs, as low drug solubility in aqueous media hampers the ability of drugs to be administered through the intravenous route. Since many important anticancer agents have poor water solubility, the development of novel delivery systems for these molecules without the use of organic solvents will be a step forward towards a better solution/s. inorganic CRT systems provide good nanotechnology platforms that will provide opportunities to develop new therapeutic agents for targeted drug delivery. It is expected that the inorganic CRT nanotechnology will make significant advances in cancer treatment in the future even though there has been excellent progress with other technologies, such as quantum dots,[14,15] nanoshells,[16] gold nanoparticles,[17] and carbon nanotubes.[18]

In general, taking the advantages of good control of molecular and/or nano-dispersion and nano-distribution of API, the inorganic CRT nanotechnology introduced in this book will lead to a new era for future drug development.

2.3 BASIC UNDERSTANDING OF POTENTIAL MOLECULAR INTERACTIONS

2.3.1 Classical API Excipients

Classical drug formulations are a combination of API and excipients. Excipients are included in dosage forms to aid manufacture, administration, or absorption and must be able to aid the API in fulfilling its functions, that is, dosage, stability, and release of API from the formulation. It is normally not good practice to consider excipients as inert as they can, if used adequately, initiate, propagate or participate in chemical or physical interactions with drug compounds to maximize the effectiveness of a medication. The definition of excipients as developed by IPEC (International Pharmaceutical

Excipients Council) America and IPEC Europe is, '*These are the substance(s) other than the API which has been appropriately evaluated for safety and is included in a drug delivery system to either aid processing of the system during manufacturing or protect, support or enhance stability, bioavailability or patients compliances or assist in product identification and enhance any other attributes of overall safety and effectiveness of drug product during storage or use*'.[19]

Excipients have assistant roles in drug formulations. There are a total of only 23 functionalities among 94 excipients that are used in 200 of the most popular tablets and drug capsules.[20] Some of the functions of excipients will be still considered when employing inorganic CRT for drug formulation and manufacturing. This is because it is common that all formulations need some excipients to be used to control *chemical stability, dissolution, solubility,* and *bioavailability* of the API.[21,22] Due to the complexity of the subject, discussions on the interaction between API and excipients are beyond the scope of this book. In reality, few definitive theories have ever been developed to describe the working mechanisms and interactions between the excipients in a formulation and the incumbent API.

However, the most important molecular interactions have been recognized. They are hydrogen bonds, van der Waals, and electrostatic interactions,[23–27] among which the hydrogen bond is the strongest and most well-known of the intermolecular forces.[28] These intermolecular interactions are coherent intermolecular forces of a physical nature, and it is these forces that lead to *cooperative interactions* between the same and/or different molecules and on which most supramolecular structures are formed. This molecular cooperative interaction has been coined as supramolecular chemistry,[26] a feature of which is the lack of any new chemical bonds being formed between the molecules or systems involved. In the current context we are only interested in intermolecular interactions in the view of molecular physics but based on a sound understanding of materials chemistry, that is, chemical structure.

2.3.2 Interactions Between API and inorganic CRT Matrix Systems

Combination of materials chemistry, physics, and inorganic CRT nanotechnology introduced in Chapter 1, this chapter and more in subsequent chapters will lead to significant advances in many areas for new drug formulation and manufacturing technology. Unlike classical drugs and excipients that have standards that have been well established in most circumstances, the use of inorganic CRT materials as novel controlled release systems is at an early stage of development. So the role of excipients in new inorganic CRT formulations will be very different as compared with classical drug formulation. Here will be introduced new concepts and fundamental understanding on the relevant cooperative interactions between API molecules and inorganic CRT matrices.

The fundamental point of molecular interactions is to understand the forces coherently associated with each of the individual molecular species involved, which in turn is based upon the basic physical forces involved. van der Waals forces are interactions between molecules without affecting their molecular structure and can be repulsive or attractive depending on the distance between the interacting non-bonded atoms and molecules. Although these forces are individually weak, say from 2 to 20 kJ mol^{-1}, compared with the 348 kJ mol^{-1} of covalent C—C bonds as shown in Table 2.1, they become significant when considered in numbers. One good example is polymers—plastics and rubbers that we use daily. It is the intermolecular forces that hold each individual macromolecule together and in some cases are strong enough to produce engineering quality materials.

Bear in mind that intermolecular interactions exist in all molecules; now we will extend this concept to organic-inorganic interactions. This is because most APIs are naturally organic based and materials employed for use in inorganic CRT materials are inorganic. We will provide an overview of the present state and upcoming challenges in this growing and highly interdisciplinary research field. It is intended to detail how these interactions of the chosen inorganic CRT materials can generate novel functional features such as the loading and release of drug molecules in a controlled manner through weak but important cooperative molecular interactions. Later in this book

Table 2.1 Typical strengths of intermolecular forces

Force	Strength (kJ mol^{-1})	Notes
Ion-Ion	~250	Only between ions
Hydrogen bonding A—H···B	~20	Applicable to: A, B = N, O or F
Ion-dipole	~15	
Dipole-dipole	~2	Between stationary polar molecules
London (dispersion)	~2	Between all molecules
Covalent C—C bonding	*~348*	*Not van der Waals; for comparison only*

Adapted from Refs. 29 & 30.

(Chapter 4), we will discuss the subject in much more detail in the two specific material groups, ordered and non-ordered mesoporous silica systems.

2.3.3 The Surface Chemistry of Silica

In the 1930s, studies on the condensation processes of silicic acids by Kiselev[31] showed that hydroxyl (silanol) groups, Si—OH, were present on the surface of silicate and silica materials. This conclusion was obtained by observation that the water evolved during calcination of silica gel is formed from OH groups that are chemically held on to the silica surface. This naturally leads to the questions: what is the concentration of silanol groups on the silica surface and where do the silanol groups come from?

Zhuravlev[32] systematically studied and subsequently calculated the concentration of surface silanol groups for a range of silica glasses. His conclusion was around 4.6 Si—OH nm^{-2} (i.e., 4.6×10^{12} Si—OH mm^{-2}), which is independent of the differences in the preparation method used and the surface porosity of the silica materials. Formation of silanol groups occurs *via* two routes. The first is as a consequence of the chemical reactions taking place between silica species during the sol-gel synthesis of the product nanoporous silica materials, as typified by the reaction scheme in Figure 1.5. Due to the random nature of the particle growth during the sol-gel process, a significant proportion of the free silanol species present will be unable to react with an adjacent silicon centre because of unfavourable steric conditions. Following the final ageing step of the sol-gel process these unreacted silanol groups would be found to reside either on the external surface, on the walls of surface accessible pores, or on the walls of non-surface accessible pores.

In addition, surface Si—OH groups can also be formed as a result of rehydroxylation from dehydroxylated silica following interaction with moisture or aqueous solutions. This is better explained in Figure 2.12—a schematic plot that highlights the processes of rehydroxylation and dehydroxylation. To reach a complete dehydroxylation status, the temperature would need to be as high as 1200 °C. However, even following that level of intense thermal treatment, as the dehydroxylation conditions are removed, an equilibrium will be set up between the silica surface and the atmosphere such that hydroxylation of at least a small percentage of the surface silicon atoms will be extremely favourable. Therefore the surface chemistry of silica would be mainly dominated by hydroxylation under most circumstances; hence Si—OH groups are the most prominent chemical features on silica surfaces.

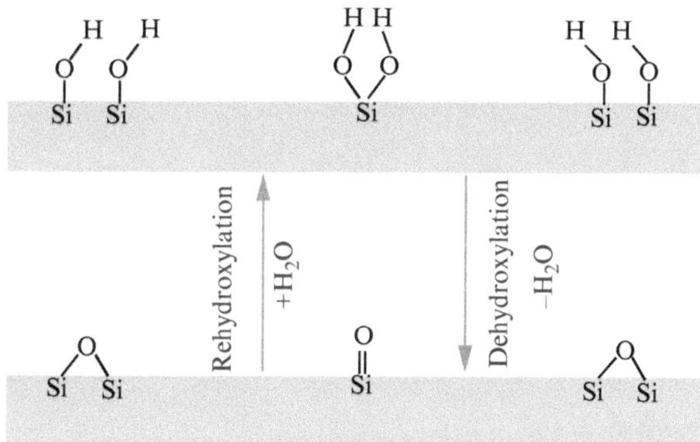

Figure 2.12 Schematic illustration of dehydroxylation and rehydroxylation on the surface of mesoporous silica.

2.3.4 Molecular Interaction with Directionally Templated Mesoporous Silica Systems

Amorphous silica has a long history of being employed in pharmaceutical formulations as an anticaking excipient. Most recently, mesoporous silica such as MCM-41 has attracted worldwide attention for its potential as a drug carrier for the controlled release of drugs. However, there is relatively little known on how the silica interacts with the adsorbed API molecules in particular, the interactions taking place at the molecular level.

Table 2.2 collects some on-market drugs as examples. By examining the drugs' chemical structures, it is not difficult to find that all 17 compounds have common structural features, that is, they all have at least one of the following groups: hydroxyl (—OH), carbonyl (>C=O), primary amine (—NH_2), or secondary amine (>NH). With reference to Table 2.1, they are the functional groups that are capable of forming hydrogen bonds, the strongest force due to intermolecular interactions, in addition to other forms of intermolecular interactions as they are generally all polar molecules and relatively large ones. Taking the fact that the concentration of Si—OH is 4.6 nm^{-2}, there will be good opportunities to have the drug functional groups, —OH, >C=O, —NH_2, and >NH, to form H-bonds in mesoporous silica pores that are formed through templated silica systems (read Chapter 4 for further details).

Let us use ibuprofen as an example to investigate the interaction between mesoporous silica MCM-41 and the drug molecules. Figure 2.13 is a computing simulation with a pore diameter around 3 nm. There are six

Table 2.2 List of drug molecules

Drug	Chemical structure
Ansamycins	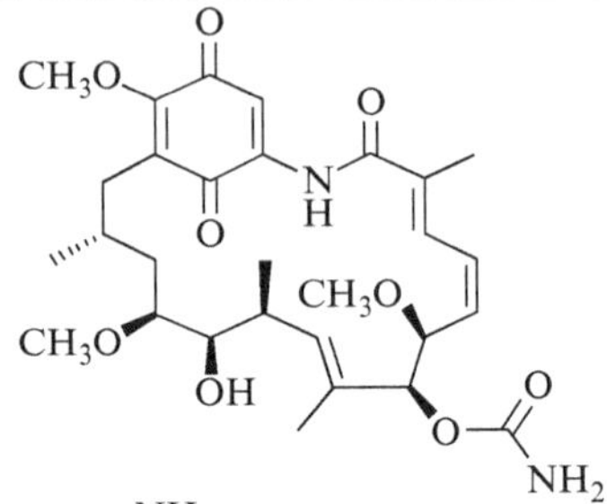
Carbacephems	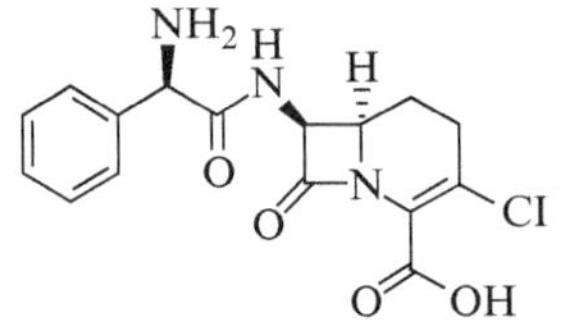
Lincosamides	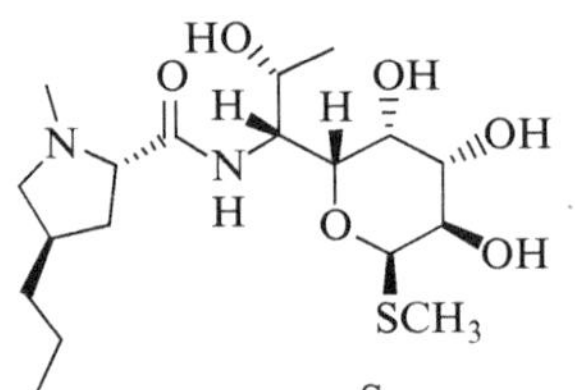
Monobactams	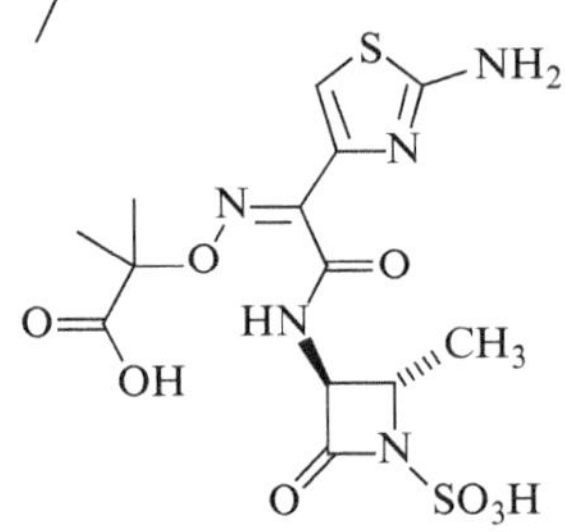
Nitrofurans	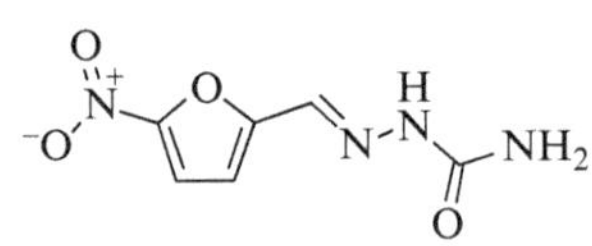
2-Oxazolidone	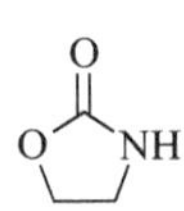
Penicillin	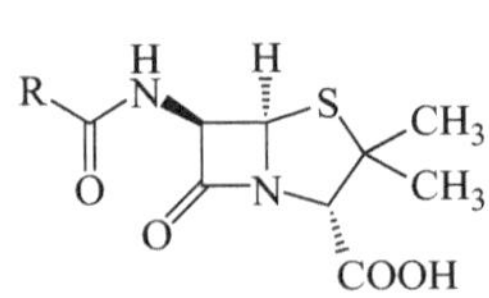

Table 2.2 List of drug molecules—cont'd

Drug	Chemical structure
Cyclandelate	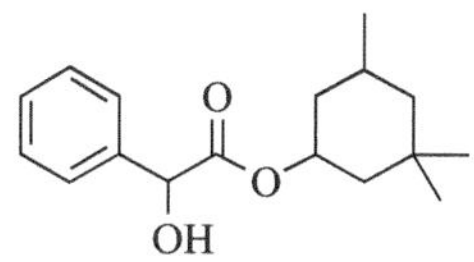
Isoxsuprine	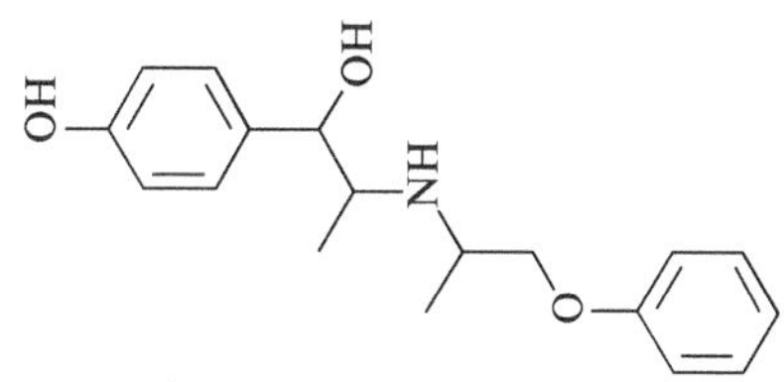
Vitamin E	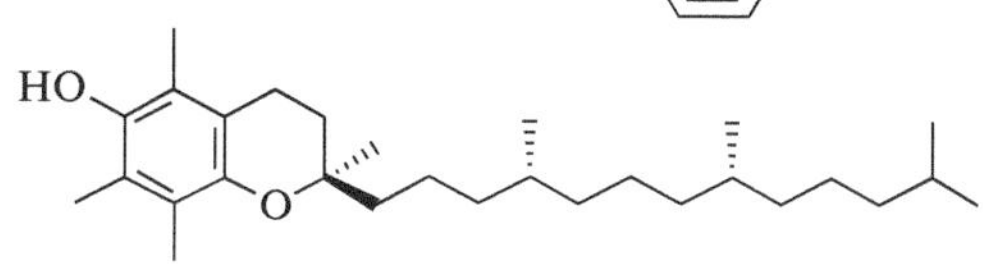
Dipyridamole	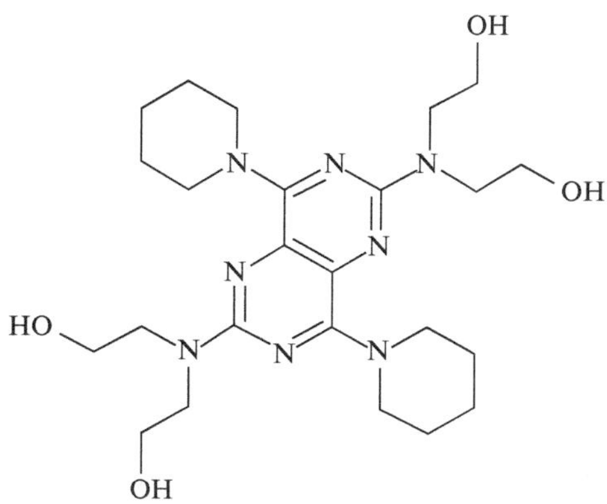
Acetylcholinesterase inhibitor	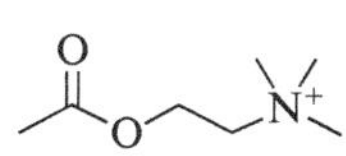
Memantine	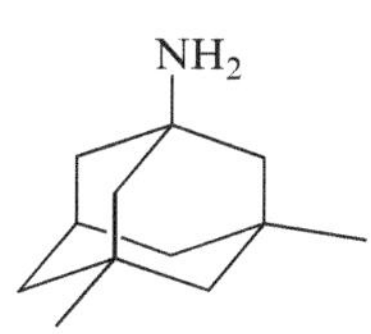
Aspirin	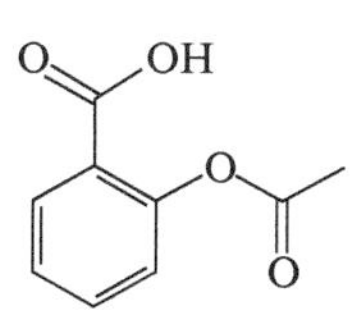

Continued

Table 2.2 List of drug molecules—cont'd

Drug	Chemical structure
Butorphanol	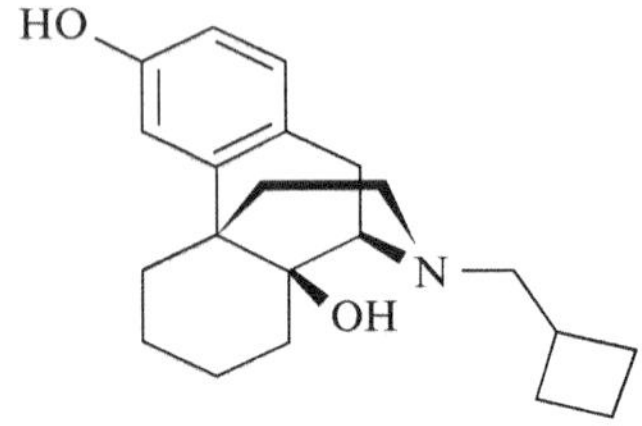
Ibuprofen	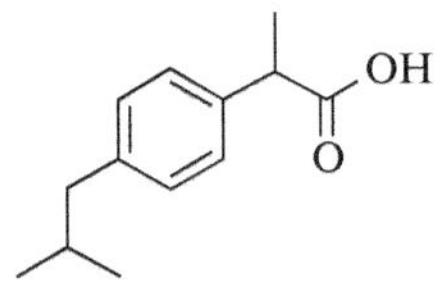
Sirolimus	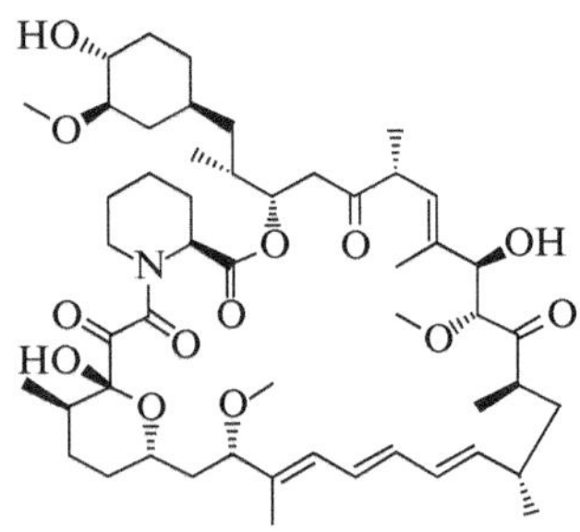

Figure 2.13 Computing simulation of molecular interaction between mesoporous silica MCM-41 and ibuprofen molecules. Deep and light grey spheres represent ibuprofen molecules and yellow (grey in the print version) and red (dark grey in the print version) represent silanol groups. *(Reproduced courtesy of Prof. Piero Ugliengo and Massimo Delle Piane; University of Turin).*

ibuprofen molecules situated inside the nanopore; H-bonding formation between Si—OH groups and the carboxylic acid group (CO_2H) of ibuprofen is triggered by the adsorption of the drug molecules onto the surface of the silica nanopores. In addition to H-bonds, other van der Waals forces such as dispersion interactions also play a crucial role in the formation of different configurations due to the drug-silica interaction. The combined interactions of directional H-bonds and non-specific dispersion interactions will lead to the success of drug loading. This computing modelling confirms, in principle, the possibility.

Experimentally, Charnay[33] employed mesoporous silica MCM-41 as a drug carrier. The particle size of the MCM-41 is about 50 μm, so it could reasonably be expected that the adsorbed drug molecules may not be able to penetrate from the surface fully into the core of the particle, and that the maximum drug loading may not be achieved. However the results confirmed that a high drug loading was achieved.

Let us analyse this silica carrier and determine the free volume present and subsequently the total volume fraction of ibuprofen loaded into the mesoporous silica particles.

Take the experimentally measured specific volume of 0.63 $cm^3\ g^{-1}$ as a base. We can work out the apparent density of MCM-41, which is 1.587 $g\ cm^{-3}$. Now use Equation (2.1) to estimate the total free volume of the silica—still use the density 4.287 $g\ cm^{-3}$ of extremely dense tetragonal form (polymorph) of silicon dioxide. The total free volume of the MCM-41 is 63% (refer to Figure 2.5 too). This means that in this study, for this particular sample, the total free volume of the mesoporous particles is very high, that is, SiO_2 molecules or a group of SiO_2 molecules only occupy 37% of the volume of the MCM-41.

Now we should compare this free volume with the measured drug volume occupied within the mesoporous silica. Taking the drug loading of 500 mg/g into account as determined experimentally[33] and by using the density 1.03 $g\ cm^{-3}$ of ibuprofen, the volume fraction V^f_{drug} or volume percentage of the ibuprofen can be determined using the following equation:

$$V^{f}_{drug} = \frac{V_{drug}}{V_{total}} = \frac{V_{drug}}{V_{drug} + V_{MCM41}} = \frac{f_{drug}\rho_{MCM41}}{f_{drug}\rho_{MCM41} + f_{MCM41}\rho_{drug}} \quad (2.2)$$

where f_{drug} and f_{MCM41} are the weight fractions of drug and silica MCM-41 in the total weight of the mixture. Alternatively, if the drug loading is based on mesoporous silica, not total weight, that is, the ratio of the weight of drug over the weight of silica, the volume fraction V_{drug}^{f} or percentage of volume of the ibuprofen can be calculated using the following equation:

$$V_{\mathrm{drug}}^{\mathrm{f}} = \frac{V_{\mathrm{drug}}}{V_{\mathrm{total}}} = \frac{V_{\mathrm{drug}}}{V_{\mathrm{drug}} + V_{\mathrm{MCM41}}} = \frac{1}{1 + \dfrac{W_{\mathrm{MCM41}}}{W_{\mathrm{drug}}} \dfrac{\rho_{\mathrm{drug}}}{\rho_{\mathrm{MCM41}}}} \quad (2.3)$$

In Charnay's report the drug loading result is expressed in 'mg/g'. It is not clear if this result is calculated based on the total weight (silica + drug) or based on the silica weight only. If it is the former, the volume percentage of ibuprofen would be 60.5% according to Equation (2.2). If it is the latter, the corresponding volume percentage would be 43.4% according to Equation (2.3). Nevertheless, it appears to be a very successful system because it has reached such a high drug loading. A large proportion of the total theoretical free volume, which is estimated to be 63%, is occupied. To achieve a molecular interaction between drug and surface of the silica, the nanopores must play a key role.

Recall the surface material chemistry of silica discussed. Strong and persistent molecular interactions are occurring between the silanol groups in the silica and the acid groups on the ibuprofen (refer to Table 2.2). The Si—OH groups are primarily taking part in hydrogen bonding interactions with the acid groups of ibuprofen. Additionally, adsorption will be strengthened by a significant contribution from the effects of van der Waals interactions between the ibuprofen molecules and the nanopores. Fundamentally both directional H-bonding and non-specific dispersion interactions operate in conjunction to enhance the interactions between the drug and the silica matrix to achieve high drug loading. This principle is not only applicable to ibuprofen, but to all other drugs such as those examples listed in Table 2.2, where almost all drugs have functional groups that can potentially form H-bonds with the Si—OH residing on the nanopore surfaces as well as general interactions between the drug molecules on the silica *via* a range of van der Waals forces.

2.3.5 Towards Molecular Dispersion and Distribution

Another important feature worth addressing here is how inorganic CRT delivery systems can achieve levels of drug dispersion on the molecular level. Classical drug formulation processing and manufacture would be unlikely to achieve API molecular dispersion and distribution within the delivery matrix with a range of different excipients. In the above case (Charnay), two significant formulation properties were achieved: high drug loading and molecular dispersion. This was confirmed by the observation that no ibuprofen crystalline structure was seen in the XRD diffractogram after producing the ibuprofen-loaded mesoporous silica. This is a significant finding that suggests that even at high drug loading levels, locating the ibuprofen molecules within the nanoporous structure, in the first instance, means that the ibuprofen has achieved molecular or at least pseudo-molecular dispersion. It also suggests that enclosing the ibuprofen molecules within the nanoporous structure prevents the nucleation of a sufficient number of ibuprofen molecules to enable crystallite formation. Not only does this property have the effect of reducing the energetic barrier to drug dissolution (compare the dissolution rates of amorphous solids vs. crystalline solids), but has additional benefits in terms of producing a formulation that has kinetically trapped the ibuprofen in a metastable (amorphous) state. This is one of the key advantages of inorganic CRT discussed in Section 1.1 in Chapter 1. This molecular dispersion and distribution of APIs increases the apparent solubility and as such their bioavailability.

2.3.6 Molecular Interaction Sites on Sol-Gel Silica and Phosphate Glass

Mesoporous silica made from directional templates has well-defined and designed nanostructure (also refer to Chapter 4). Sol-gel- and fusion-derived glasses, two other inorganic CRT systems introduced in Chapter 1, cannot achieve such high levels of regulated nanostructure. However, they have other unique advantages over the directional templated silica, a simple benefit being that these materials require no other reagents apart from those which are present in the final glass products. This results in a significant reduction in the potential for process contamination being introduced. The manufacturing process is also much simpler, and there is also additional flexibility for changing the glass properties to enable efficient incorporation of drug molecules, as shown in Figures 1.3, 1.4 and 1.6 in Chapter 1.

The subject of API-matrix interactions has not been explored at all with phosphate glassy materials, while the equivalent work with sol-gel silica glasses has only recently begun to be studied. As such the ideas and theories used to model the API-matrix interaction mechanisms at the molecular level with respect to controlled release of drug, therapeutic ions, and other medicinal and non-medicinal active ingredients is still very much in its infancy. Here we will introduce some basic concepts to help to understand the subject.

Let us start by reviewing the basic principles of glass surface chemistry, the fundamentals of which will aid the understanding of the overall science and their potential applications. At the molecular level, glasses are composed entirely of oxides. This is worth noting because any interactions between drug functional groups and glass surfaces should be relevant to the surface chemistry and materials physics of the specific surface functionalities involved. These interactions must be considered at all stages of a new inorganic CRT-drug system development process, from the raw glass precursors to the final manufactured formulation.

In general, sol-gel silica glass has the same SiO_2 chemistry as was discussed previously with the directional templated mesoporous silica and as such the interactions of APIs with the sol-gel silica matrix should be very similar to that which has already been discussed in the context of surface silanol chemistry (Si—OH). This group has great potential to be developed further to introduce new and enhance existing modes of interaction mechanisms. This is typically carried out by manipulating the reactivity of the Si—OH group to introduce other functional groups such as NH_2, SH or CO_2H. The choice of functional group should largely be dictated by the functional groups present in the drug requiring formulation. In this way it can be seen that inorganic CRT delivery systems have the flexibility to encompass a range of APIs with varying properties. This subject will be discussed in more detail in Chapter 5 (Section 5.4).

Recently the potential use of phosphate glasses (P-glasses) for biomedical applications has stimulated considerable extra research. These materials have many favourable characteristics as compared with the properties of silicate glasses. For example, their compositions can be designed and formulated such that they can be dissolved and/or degraded into basic constituents that are biocompatible and in fact well tolerated in the body. Their dissolution rate can be designed to span over several orders of magnitude. This property is a particular benefit of P-glasses which is why these materials are increasingly being seen as ideal candidates for application as inorganic CRT formulations to achieve improved drug and ion release. A more detailed discussion of iCRT formulations releasing therapeutic ions such as Li, Ca, Cu, Sr, Mg, Zr, Ce, Ga, Co, as well as other active ingredients, will be discussed in Chapters 5–7.

Again, understanding the chemical properties of P-glasses is crucial in being able to prepare novel materials with the desired functionality. Table 2.3 collates 27 oxides that are often used for glass formulation and manufacturing, including P-glass. The table also contains cation radius (r)

Table 2.3 Commonly used oxides and their corresponding cationic radii and normalized charges

Oxide	Ion	*r*	+1/*r*	Oxide	Ion	*r*	+1/*r*	Oxide	Ion	*r*	+1/*r*	Oxide	Ion	*r*	+1/*r*	Oxide	Ion	*r*	+1/*r*
Li_2O	Li^+	76	13.2	BeO	Be^{2+}	45	44.4	B_2O_3	B^{3+}	27	111.1	SiO_2	Si^{4+}	40	100.0	P_2O_5	P^{5+}	38	131.6
Na_2O	Na^+	102	9.8	MgO	Mg^{2+}	72	27.8	Al_2O_3	Al^{3+}	53	56.6	GeO_2	Ge^{4+}	53	75.5	As_2O_5	As^{5+}	46	108.7
K_2O	K^+	138	7.2	ZnO	Zn^{2+}	74	27.0	Ga_2O_3	Ga^{2+}	62	48.4	TiO_2	Ti^{4+}	67	59.7	Sb_2O_5	Sb^{5+}	60	83.3
Rb_2O	Rb^+	152	6.6	MnO	Mn^{2+}	83	24.1	As_2O_3	As^{3+}	58	51.7	ZrO_2	Zr^{4+}	72	55.6				
Cs_2O	Cs^+	167	6.0	PdO	Pd^{2+}	86	23.3	Sb_2O_3	Sb^{3+}	76	39.5								
				CaO	Ca^{2+}	100	20.0	Bi_2O_3	Bi^{3+}	103	29.1								
				SrO	Sr^{2+}	118	16.9	P_2O_3	P^{3+}	44	68.2								
				BaO	Ba^{2+}	135	14.8												

Notes: ion radius *r* is in pm; normalised $+n/r$ is in $+n\text{pm}^{-1} \times 10^3$; the radii are the values assuming, having six coordination number in form for comparison purpose based on the same standard. Raw data: http://abulafia.mt.ic.ac.uk/shannon/ptable.php.

measured in picometres (pm). Further treatment of the data is required to normalize the charge with respect to the corresponding ionic radius giving the normalized charge '$+n/r$', that is, charge per unit length, in pm^{-1}, where 'n' is the respective cationic charge. To the first approximation, the normalized charge of a cation can be treated as a one-dimensional charge density, meaning the average charge per unit length; it can also serve as the positive charge potential for these cations. This charge density or potential is useful in designing P-glass formulations (as well as other types of glasses) and for predicting product dissolution rates. It is also a good indicator of the potential for activated sites being present on the surface which are common due to defects and/or dislocation of negatively and positively charged ions.

Figure 2.14 shows the normalized charge potentials obtained from Table 2.3, plotted as a function of decreasing cationic charge. These cations and their corresponding charge potential can be roughly divided into different regions according to their roles in glass formulations, based on the concepts introduced in Chapter 1 (Table 1.2):

(1) network former,
(2) intermediate,
(3) network modifier (consisting of a range of oxides from moderate to severe network disrupters).

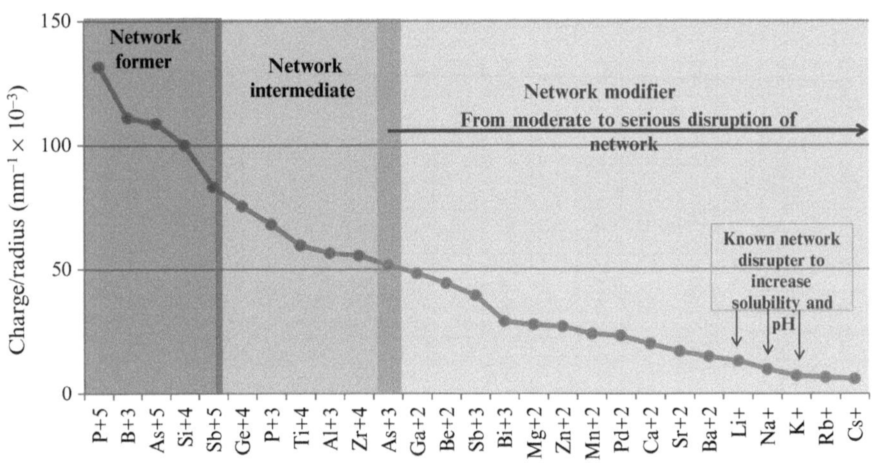

Figure 2.14 Normalized cation charge of different ions.

From this analysis of the materials chemistry, we will be able to apply more fundamental principles to help interpret the physical meaning of their properties and predict possible molecular interactions.

The basic principles and interaction mechanisms between APIs and P-glasses should be similar to that as discussed for silica glass. However, there are other additional factors that could affect interactions at the molecular level. What is the chemical structure of a P-glass surface? There are always positively and negatively charged ions on the surface due to the presence of defects and dislocations. This is an area where there has been relatively little research carried out, particularly in the context of relating the surface chemistry to the controlled release properties of the materials.

Let us use one case study to evaluate not only the material chemistry but also the mechanisms involved with the effects that are being seen.

Case study (refer to Chapter 1 on fused glass technology):

(1) A water-soluble P-glass formulation: P_2O_5: 40 mol%; CaO: 25 mol%; Na_2O: 5 mol%; MgO: 30 mol%.
(2) Making the P-glass: place the oxide precursors in a quartz crucible and heat to 1200 °C for 1 h, quench the molten glass in distilled water.
(3) Postprocessing: grind/mill the glass made in (b); sieve and collect glass particles with the required particle size distribution; wash with ethanol and dry in an oven at 80 °C.
(4) Characterization: Surface chemistry on the top 1-2 nm using ToF-SIMS (time-of-flight secondary ion mass spectrometry, more characterization methods in Chapter 3).

By employing ToF-SIMS analysis, secondary ions were detected from just the very top surface (1-2 nm) to give an indication of the molecular groups present (See Section 3.2.5). They are listed as follows:

(1) positively charged ions: Na, Mg, MgOH, Ca, CaO, CaOH.
(2) negatively charged ions: PO, PO_2, PO_3, PO_4H, PO_4H_2, O, OH.

These additional ionic species are included in Figure 2.14 to give Figure 2.15.

It is known that phosphateis four-coordinate with oxygen atoms at the corners of a tetrahedron. The modification from one phosphate species to another can be described in terms of the number of bridging oxygens present within each phosphate tetrahedron.[34] In line with other glass families, the notation Q^n is used to describe the extent of network connectivity within phosphate glasses where n = the number of bridging oxygens present.

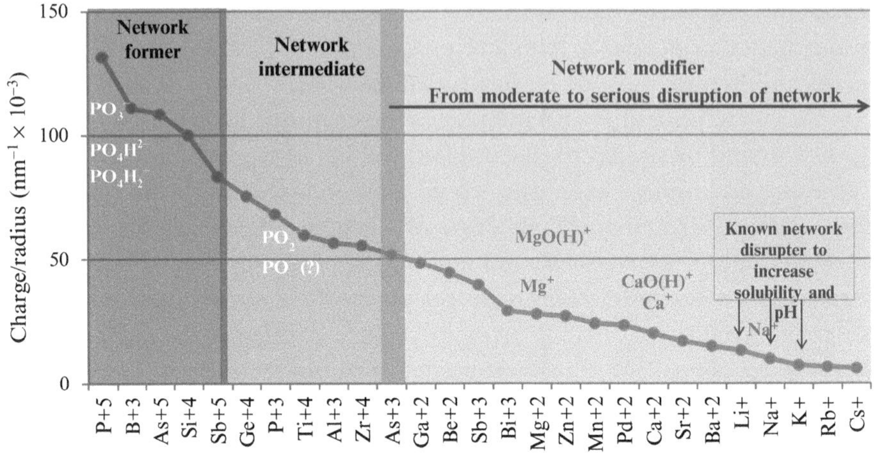

Figure 2.15 Normalized cation charge of oxides with ionized clusters obtained from the case study in Section 2.3.6.

It can now be seen that a distribution of phosphate species with different numbers of bridging oxygens can be produced depending on the chemical nature of the network modifier ions present and in what molar amount. By considering the composition of the P-glass above, the network connectivity of the bulk glass can be theoretically determined which in turn can be an indicator of the distribution of the Q^{0-3} network forming species present (analytical techniques such as FTIR can indicate the relative amounts of Q^n species actually present in the bulk glass – see Section 3.2.3). Considering the theoretical network connectivity, and the associated number of non-bridging oxygen atoms present as a result, can provide guidance as to the relative numbers of bridging, non-bridging and double bond oxygens present at the P-glass surface, each of which could serve as distinctly different active sites for API interaction with P-glasses at the molecular level.

In this respect, if the surfaces of phosphate glasses are chemically analogous to the surfaces of silicate glasses, rehydroxylation of the phosphate glass surface could occur, as discussed in Section 2.3.3 (also refer to Figure 2.12). If so, on the surfaces of P-glasses there should be many molecularly active sites that can potentially serve as drug interaction points. However the mechanisms of interaction will be more complicated than the pure silica glass systems since phosphate glasses are necessarily formulations of mixed oxide systems. Again these new systems create more questions than answers that are worth further investigation. Principally, what is the best way to enhance the interactions between different drugs and hybrid P-glass systems in order to design improved drug delivery systems through fundamental understanding of the surface chemistry and physics?

2.3.7 Dissolution of Phosphate Glass

The case study of P-glass discussed in Section 2.3.6 has shown how a consideration of phosphate glass composition has led to discussions on the relevant theories concerning the glass structure. The distribution of different amounts of Q^n phosphate species can serve as an indicator of the relative amounts of bridging and non-bridging oxygens present in the bulk glass and at the surface. The presence of associated network modifying cations with different charge strengths (as seen on the right in Figure 2.15) will have an impact on the relative rates of cation dissolution when placed in aqueous media. This in turn will directly affect the rate of drug release from a drug-loaded P-glass.

Case study on P-glass dissolution: The same material used for the study in Section 2.3.6 is also used here. Evaluation of the dissolution of the P-glass was carried out in aqueous media. The particle size was controlled to have a distribution between 250 and 500 μm. Figure 2.16[35] shows the dissolution results: the normalized dissolution concentration of Na, Ca, and Mg ions as a function of time. Based on the theory of normalized charge density as introduced earlier, it is predicted that the dissolution speed should be in the order of:

$$Na^+ > Ca^{2+} > Mg^{2+}$$

This is the same order of the normalized charge strength of the three cations in Table 2.3. The results shown in Figure 2.16 confirm this prediction.

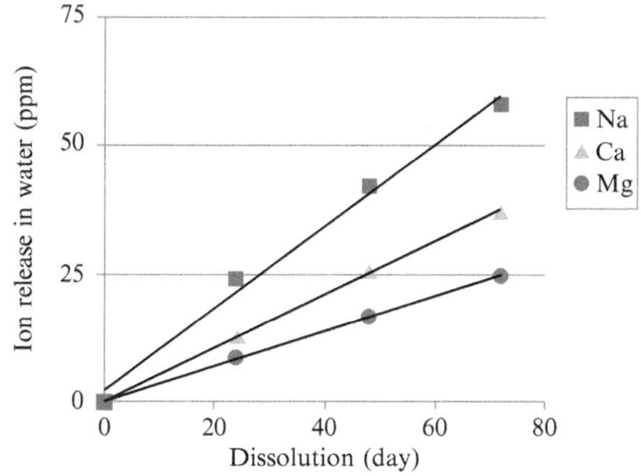

Figure 2.16 Normalized cation dissolution rates from a P-glass.

In general, P-glasses are formulated to have varying levels of solubility in aqueous media. The dissolution behaviour, or rate of dissolution, is of course dependent on the P-glass composition, and the flexibility in designing these materials exists to enable the preparation of both very rapidly dissolving and very slowly dissolving P-glasses.

Figure 2.14 can serve as a general roadmap for the production of P-glass formulations to alter their solubility. It is predicted that increasing P-glass solubility can be achieved by employing increasing mol% of the oxides on the right in the figure. The theory being that changing the formulation composition can not only affect the phosphorus oxygen status, that is, from a state of bridging oxygen atom to non-bridging, but also by weakening the cation charge potential more when the formulation consists of more oxides from the right side, which are known network interrupters such as Li_2O, Na_2O and K_2O.

Figure 2.17 shows the controlled release of ibuprofen[36] from a water-soluble glass that was made with the same formulation but by including three different modifier oxides: Li_2O, Na_2O, and K_2O. In this case it would be expected that the releasing rate of ibuprofen should be in the order of $K_2O > Na_2O > Li_2O$ of the water-soluble glasses. This is because the normalized cationic charge strength (refer to Table 2.3) is

$$K^+ (7.25) < Na^+ (9.80) < Li^+ (13.16) \left(\text{unit in charge pm}^{-1} \times 10^3\right)$$

Bear in mind that the cation radii in Table 2.3 are assumed to have a coordination number of 6, based on which the normalized charge strength

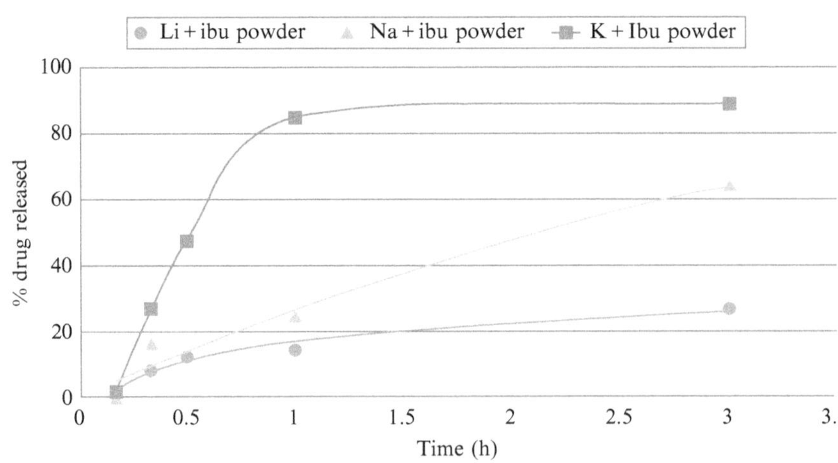

Figure 2.17 Ibuprofen controlled release from different soluble glasses.

per unit length was calculated. This is not a definitive trend, and it is expected that there are exceptions as there are many other factors that can affect both the ionic charge density and ionic radius from which this trend is derived. Figure 2.14 therefore only serves as a general guideline for new glass formulation design and for developing inorganic CRT systems for controlled drug release and other applications.

2.3.8 Glass Formulation for inorganic CRT

The most commonly used formulations for both sol-gel- and fusion-derived silicate glasses contain three oxides: CaO, SiO_2 and P_2O_5. The CaO-SiO_2-P_2O_5 system is an important one for inorganic CRT technology development. Figure 2.18 is a summary that provides researchers a formulation roadmap.[6] It is a general guideline for the design and development of sol-gel and fused glasses, which are highlighted in the figure. Some compositional ranges are suitable for sol-gel processing only and equally some are suitable only for fusion process, while others still are amenable to both processes. The formulations highlighted are a mixture of silicate-based and

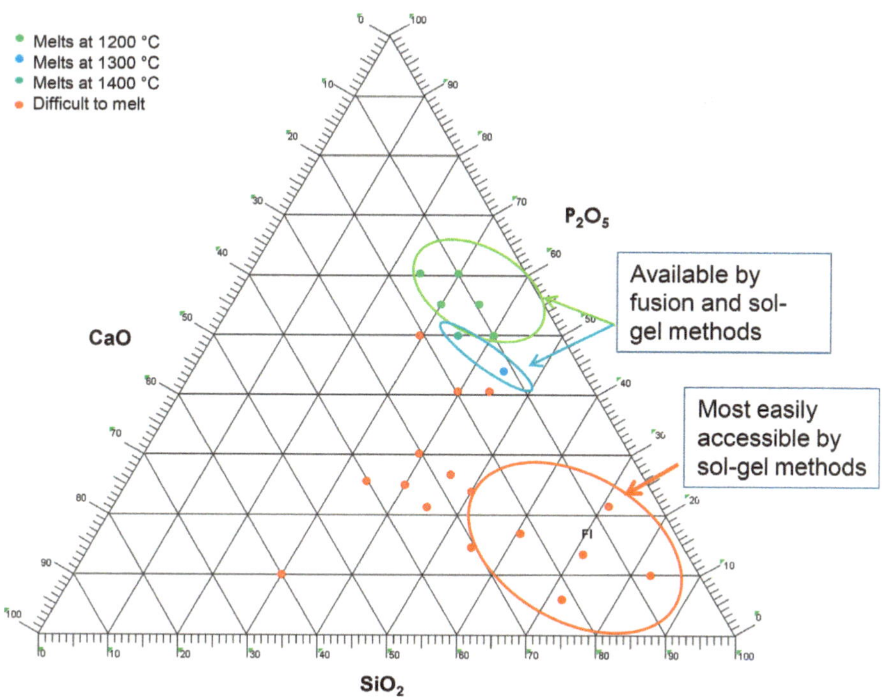

Figure 2.18 Roadmap for formulation of CaO-SiO_2-P_2O_5 glasses.

phosphate-based materials. The formulation roadmap when combined with all other oxides listed in Table 2.3 and Figure 2.14 will provide researchers with a range of choices to help design and develop new inorganic CRT systems for controlled drug release and other applications.

2.4 THEORY AND PRACTICAL MODELLING OF DRUG CONTROLLED RELEASE KINETICS

Theory and practical modelling of controlled release kinetics is important and can help to elucidate the fundamental processes and mechanisms involved during controlled drug release from inorganic CRT systems. There are many research and review papers discussing the subject area.[37,38] This book will not attempt to review the entire field but will highlight and discuss some of the kinetic principles based on the original theoretical and practical models used by many previous and current researchers. By doing so, readers will not only have practical models to aid their study but the theories and *associated assumptions and boundary conditions* for the practical models often used in publications.

When a mesoporous structure is loaded with drug molecules, such as that shown in Figure 2.13, the formulation will firstly be brought into contact with the chosen model dissolution fluid, following which a series of mass transport phenomena will occur:

(a) A nanopore in a mesoporous system will be filled with water; this is a diffusion-controlled process.

(b) Drug molecules will be dissolved in the water-filled nanopore; this is a dissolution process.

(c) The dissolved drug molecules will diffuse out of the nanopore; this is also a diffusion-controlled process.

The situation, however, is not quite that simple. The complexity is added by the variation of the diffusion and dissolution rates of (a), (b), and (c), including the following possible combinations:

(1-1) (a) > (b)
(1-2) (a) = (b)
(1-3) (a) < (b)
(2-1) (b) > (c)
(2-2) (b) = (c)
(2-3) (b) < (c)
(3-1) (c) > (a)

(3-2) (c) = (a)
(3-3) (c) < (a)

These kinetic variations will affect the overall drug release.

To start, we will consider a one-dimensional diffusion mechanism. This is because the mathematics of two- and three-dimensional diffusion is far more complicated and is beyond the scope of this text. Additionally, the kinetic models obtained from one-dimensional theory are enough to cover the kinetic models often used by most researchers.

Figure 2.19 shows a one-dimensional nanopore model, within which a drug is located. The model schematically shows the two processes:

- Step 1: Water molecules diffuse into the drug-loaded nanopore.
- Step 2: Drug molecules diffuse out of the nanopore with the assistance of water molecules.

With reference to the nanopore shown in Figure 2.19, both Step I and Step II are a mass transportation process *via* diffusion. The mass transportation kinetics can be described by Fick's Law. To make the analysis even simpler, we assume that Step II will not start until the completion of Step I. By doing so, we will make the mathematical derivation simpler. In reality, it will not affect the conclusions obtained toward the end of the analysis.

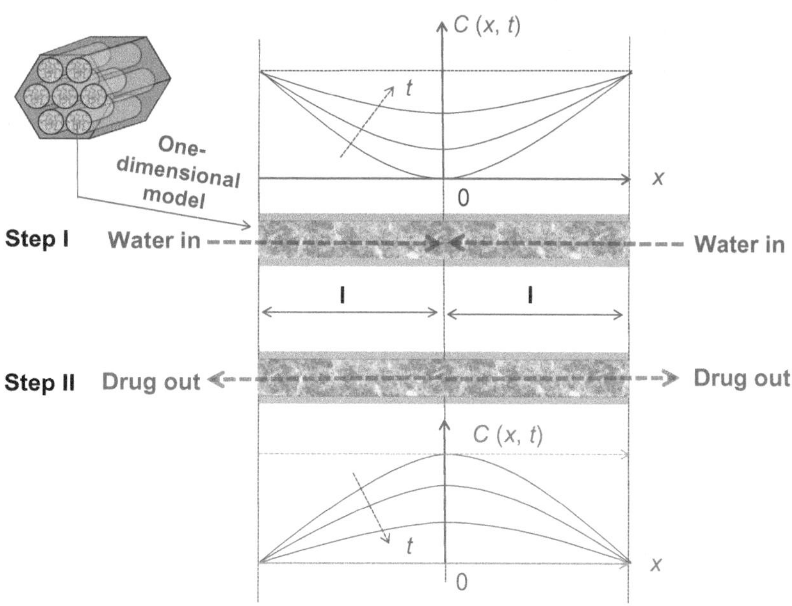

Figure 2.19 Molecular diffusion model for drug-loaded mesoporous inorganic CRT system.

The first quantitative diffusion equation was proposed by Fick in 1855. He adopted the mathematical equation of heat conduction derived by Fourier in 1822. The basic hypothesis is that the transfer rate of a diffusing substance through a unit area of a section is proportional to the concentration gradient normal to the section. This can be expressed by

$$F = -D\frac{\partial C}{\partial x} \tag{2.4}$$

where F is the transfer rate per unit area, C is the concentration of diffusion substance, x is the space coordinate normal to the section, and D is the diffusion coefficient.

Considering the one-dimensional diffusion of Step I, the change of diffusion substance C within a time '∂t' in unit area from 'x' to '$x+\partial x$' can be expressed by

$$\frac{\partial C}{\partial t} = -\left(\frac{\partial F}{\partial x}\right) \tag{2.5}$$

Assuming a constant diffusion coefficient D, combination of Equations (2.4) and (2.5) makes

$$\frac{\partial C}{\partial t} = D\left(\frac{\partial^2 C}{\partial^2 x}\right) \tag{2.6}$$

Equations (2.4) and (2.6) are known as Fick's first and second laws.

For Step I with reference to Figure 2.19a, one solution of Equation (2.6) for the one-dimensional case is given by Crank[39]:

$$\frac{C - C_i}{C_0 - C_i} = 1 - \frac{4}{\pi}\sum\frac{(-1)^n}{2n+1}\mathrm{e}^{-D(2n+1)^2\pi^2\frac{t}{4l^2}}\cos\frac{(2n+1)\pi x}{2l} \tag{2.7}$$

where C_i is the initial concentration in the region $-l < x < l$ and C_0 the concentration at $x = \pm l$ (note: this equation can be applied to both adsorption and desorption processes for Step I).

Under the conditions at $t=0$, $C_i=0$, and $C_0=C_s$, where C_s is assumed to be the saturated concentration, integrating Equation (2.7) from $x=-l$ to $x=l$ at a time t gives

$$\frac{M_t}{M_s} = 1 - \sum\frac{8}{(2n+1)^2\pi^2}\mathrm{e}^{-D(2n+1)^2\pi^2\frac{t}{4l^2}} \tag{2.8}$$

where M_t is the total amount of diffused substance which has entered into the nanopore at time t, and M_s is the saturated diffusion substance.

For the desorption process, at $t=0$, $C_i=C_s$, and $C_0=0$, where C_0 is at $x=\pm l$ immediately outside of the nanopore, integrating Equation (2.7) from $x=-l$ to $x=l$ at a time t gives:

$$\frac{M_t}{M_s}=\sum\frac{8}{(2n+1)^2\pi^2}e^{-D(2n+1)^2\pi^2\frac{t}{4l^2}} \tag{2.9}$$

It can be seen that Equation (2.9) can be obtained by subtracting Equation (2.8) from 1.

It is not difficult using a computer to work out the amount of diffused substance at a time t if the diffusion coefficient D and saturated substance M_s are known.

Diffusion coefficient D is normally obtained by a short-term diffusion test. For short diffusion times, Crank also gave different solutions to Equations (2.7) and (2.8) as follows:

$$\frac{C-C_0}{C_s-C_0}=\sum(-1)^n\mathrm{erfc}\left\{\frac{(2n+1)l-x}{2\sqrt{Dt}}\right\}+\sum(-1)^n\mathrm{erfc}\left\{\frac{(2n+1)l+x}{2\sqrt{Dt}}\right\} \tag{2.10}$$

$$\frac{M_t}{M_s}=2\sqrt{\frac{Dt}{l^2}}\left[\pi^{-0.5}+2\sum(-1)^n\mathrm{erfc}\left\{\frac{nl}{\sqrt{Dt}}\right\}\right] \tag{2.11}$$

where 'erfc' stands for error function and the rest have the same meanings as defined previously.

Normally, the first right-hand term in Equation (2.11) is used to obtain the diffusion coefficient D at an early stage of the adsorption and desorption test as the rest of the terms are negligible. This gives:

$$\frac{M_t}{M_s}=2\left(\frac{Dt}{\pi l^2}\right)^{1/2} \tag{2.12}$$

For a given system, Equation (2.12) can be written in a general form:

$$\frac{M_t}{M_s}=kt^n \tag{2.13}$$

$$\log\left(\frac{M_t}{M_s}\right)=n\log(t)+\log(k) \tag{2.14}$$

where k and n are constants. There are three cases of diffusion that can be generally classified:

- Case I: $n=0.5$, Fickian diffusion, also called Higuchi model;
- Case II: $0.5<n<1$, non-Fickian diffusion;
- Case III: $n=1$, zero-order diffusion.

Table 2.4 Ibuprofen release from MCM-41 in SBF at 37 °C

Time (h)	M_t/M_s	Time (h)	M_t/M_s
0.5	0.032	52.8	0.437
1.0	0.047	70.9	0.509
2.0	0.067	140.2	0.668
3.5	0.094	293.7	0.870
7.5	0.142	360.3	0.939
29.1	0.306	405.3	0.976

Table 2.4 shows the data associated with ibuprofen release from MCM-41. 41% (wt) drug was loaded into the mesoporous silica. This is the same set of data used in Figure 4.16, where ibuprofen release was plotted against the time$^{1/2}$. The drug release study was carried out in SBF at 37 °C. The linearity is seen up to about 50% when plotting ibuprofen release (%) as a function of time$^{1/2}$. Now we want to use the data to evaluate the drug release kinetics, that is, mechanisms.

According to Equation (2.14), Figure 2.20 shows logarithmic ibuprofen release (M_t/M_s) against logarithmic time. The slope is shown in the inset of the plot. It is 0.5638, that is, $n=0.5638$. The diffusion mechanism is neither Fickian nor zero order but is very close to the Fickian diffusion model because it is closer to $n=0.5$ than to $n=1$ of zero-order diffusion.

This section has used the simplest one-dimensional model to first reach two complete solutions of Fick's theory, which are Equation (2.8) for

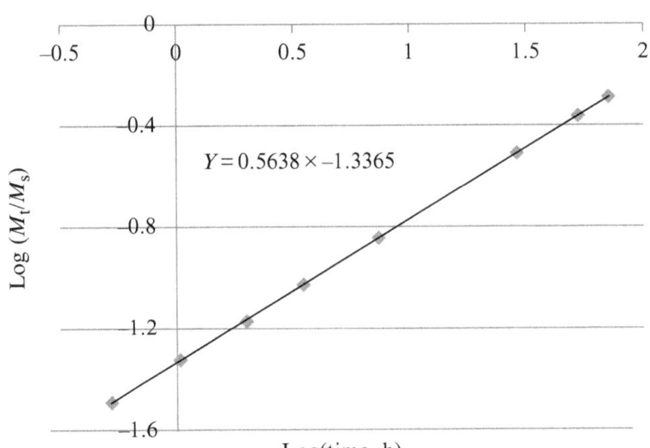

Figure 2.20 A plot of logarithmic ibuprofen release against logarithmic time.

absorption and Equation (2.9) for desorption. Then we have a simpler solution of Fick's theory, that is, Equation (2.12), which was transformed to a general model Equation (2.13). The latter is the model that is being used by most researchers. One point that requires addressing here again is that Equation (2.13) is only applicable to early-stage diffusion, typically 60% active release is the accepted upper limit depending on the system under investigation. However, the most important aspect is that we can use the equation to gain some information on the diffusion kinetics involved during the drug releasing mechanism. If it is Fickian diffusion, or zero order or something between the two by seeing the diffusion index $n=0.5$, or $n=1$ or a value between 0.5 and 1 respectively.

As addressed at the beginning of this section, there are many factors that will affect drug release kinetics. We have only considered a one-dimensional model here, and there are certainly limitations to using the model to evaluate the actual reality of the diffusion-driven processes. So there are more questions than answers to the solutions of Fick's first and second laws. This is particularly true when considering drug interaction with a mesoporous structure, drug dissolution and diffusion are occurring *via* a much more complicated geometry than the one-dimensional model illustrated in Figure 2.19. In this instance, a term needs to be included in the model which takes into account the tortuosity (τ) of the mean diffusion path length. The complexity will increase if the inorganic CRT formulation is a water-soluble system rather than one similar to mesoporous silica such as MCM-41, which compared to the typical drug release times are insoluble. In this instance dissolution of the carrier matrix needs to be accounted for in conjunction with the diffusion-mediated processes. All these additional models are beyond the scope of this book, and it is suggested that researchers with an in-depth working knowledge modelling such drug delivery systems would be better suited to establish improved kinetic models and solutions for new inorganic CRT formulations.

REFERENCES

1. Novoselov S, Jiang D, Schedin F, Booth TJ, Khotkevich VV, Morozov SV, et al. Two dimensional atomic crystals. *Proc Natl Acad Sci U S A* 2005;**102**(30):10451–3.
2. Ferrari AC. Science and technology roadmap for graphene, related two-dimensional crystals, and hybrid systems. *Nanoscale* 2015;**11**(7):4598–810.
3. Lee K, Mazare A, Schmuki P. One-dimensional titanium dioxide nanomaterials: nanotubes. *Chem Rev* 2014;**114**(19):9385–454.
4. Kauzmann W. The nature of the glassy state and the behavior of liquids at low temperatures. *Chem Rev* 1948;**43**(2):219–56.

5. Harbir K. Processing technologies for pharmaceutical tablets: a review. *Int Res J Pharm* 2012;**3**(7):20–3.
6. Zhang X. Presentation at BIO Europe Spring, Italy; March 2011.
7. Chen H, Gu J, Shi J, Liu Z, Gao J, Ruan M, et al. A composite surfactant route for the synthesis of thermally stable and hierarchically porous zirconia with a nanocrystallized framework. *Adv Mater* 2005;**17**(16):2010–4.
8. Zhao D, Yang P, Chmelka BF, Stucky GD. Multiphase assembly of mesoporous-macroporous membranes. *Chem Mater* 1999;**11**:1174–8.
9. Vallet-Regí M, Ruiz-González L, Izquierdo-Barba I, González-Calbet JM. Revisiting silica based ordered mesoporous materials: medical applications. *J Mater Chem* 2006;**16**:26–31.
10. Wu C, Chang J. Mesoporous bioactive glasses: structure characteristics, drug/growth factor delivery and bone regeneration application. *Interface Focus* 2012;**2**:292–306.
11. Hu L, Sun H, Zhao Q, Han N, Bai L, Wang Y, et al. Multilayer encapsulated mesoporous silica nanospheres as an oral sustained drug delivery system for the poorly water-soluble drug felodipine. *Mater Sci Eng C* 2015;**47**:313–24.
12. Ahern RJ, Hanrahan JP, Tobin JM, Ryan KB, Crean AM. Comparison of fenofibrate—mesoporous silica drug-loading processes for enhanced drug delivery. *Eur J Pharm Sci* 2013;**50**:400–9.
13. Lu J, Liong M, Zink JI, Tamanoi F. Mesoporous silica nanoparticles as a delivery system for hydrophobic anticancer drugs. *Small* 2007;**3**(8):1341–6.
14. Bruchez Jr M, Moronne M, Gin P, Weiss S, Alivisatos AP. Semiconductor nanocrystals as fluorescent biological labels. *Science* 1998;**281**:2013–6.
15. Seydel C. Quantum dots get wet. *Science* 2003;**300**:80–1.
16. Loo C, Lin A, Hirsch L, Lee M, Barton J, Halas N, et al. Nanoshell-enabled photonics based imaging and therapy of cancer. *Technol Cancer Res Treat* 2004;**3**(1):33–40.
17. Paciotti GF, Myer L, Weinreich D, Goia D, Pavel N, McLaughlin RE, et al. Colloidal gold: a novel nanoparticle vector for tumor directed drug delivery. *Drug Deliv* 2004;**11**(3):169–83.
18. Bianco A, Kostarelos K, Partidos CD, Prato M. Biomedical applications of functionalised carbon nanotubes. *Chem Commun* 2005;**5**:571–7.
19. Blecher L. Excipients—the important components. *Pharm Process* 1995;**12**(1):6–7.
20. Rutesh DH. *Drug topics*; 2008.
21. Brittain. *Polymorphism in pharmaceutical solids*. New York: Marcel Dekker; 1999.
22. Byrn SR, Pfeiffer RR, Stephenson G, Grant DJW, Gleason WB. *Solid-state chemistry of drugs*. West Lafayette, IN: SSCI; 1999.
23. Etter MC, Macdonald JC, Bernstein J. Graph-set analysis of hydrogen-bond patterns in organic crystals. *Acta Crystallogr B* 1990;**46**:256–62.
24. Etter MC, Reutzel SM. Hydrogen-bond directed cocrystallization and molecular recognition properties of acyclic imides. *J Am Chem Soc* 1991;**113**:2586–98.
25. Desiraju GR. Crystal engineering: outlook and prospects. *Curr Sci* 2001;**81**:1038–42.
26. Desiraju GR. Chemistry beyond the molecule. *Nature* 2001;**412**:397–400.
27. Moulton B, Zaworotko MJ. From molecules to crystal engineering: supramolecular isomerism and polymorphism in network solids. *Chem Rev* 2001;**101**:1629–58.
28. Desiraju GR. Hydrogen bridges in crystal engineering: interactions without borders. *Accounts Chem Res* 2002;**35**:565–73.
29. Atkins P, De Paula J. Atkins' Physical Chemistry 6th ed. Oxford: Oxford University Press; 2006.
30. Bond Lengths and Energies, http://www.science.uwaterloo.ca/~cchieh/cact/c120/bondel.html, accessed: 25/3/2015.
31. Kiselev AV. *Kolloidn Zh* 1936;**2**:17.

32. Zhuravlev LT. The surface chemistry of amorphous silica. Zhuravlev model. *Colloid Surf A* 2000;**173**(1-3):1–38.
33. Charnay C, Bégu S, Tourné-Péteilh C, Nicole L, Lerner DA, Devoisselle JM. Inclusion of ibuprofen in mesoporous templated silica: drug loading and release property. *Eur J Pharm Biopharm* 2004;**57**(3):533–40.
34. Gresch R, Muller-Warmuth W. X-ray photoelectron spectroscopy of sodium phosphate glasses. *J Non-Cryst Solids* 1979;**34**:127–36.
35. Stefanic M, Zhang X. Unpublished results from a Lucideon EU FP7 project. EU funding is acknowledged.
36. Zhang X. Presentation at bio international convention. Washington; 27-30 June 2011.
37. Maria G, Berger D, Nastase S, Luta I. Kinetic studies on the irinotecan release based on structural properties of functionalized mesoporous-silica supports. *Microporous Mesoporous Mater* 2012;**149**:25–35.
38. López-Goerne TM, López-García MG, Rodríguez Grada G, Ortiz Pérez I, Gómez López E, Alvarez Lemus MA. Obtaining of sol-gel ketorolac-silica nanoparticles characterization and drug release kinetics. *J Nanomater* 2013, Article ID 450483.
39. Crank J. *The mathematics of diffusion*. Oxford: Clarendon Press; 1958.

FURTHER READING

Elliott SR. *Physics of amorphous materials*. 2nd ed. Longman Scientific and Technical; 1990.

CHAPTER 3

Materials Characterization of Inorganic Controlled Release

Contents

3.1 Introduction 58
3.2 Chemical Analysis 59
3.2.1 X-Ray fluorescence 59
3.2.1.1 Case study: Contamination investigation *60*
3.2.2 Inductively coupled plasma mass spectrometry 60
3.2.2.1 Case study: Controlled release of strontium from P-glass *61*
3.2.2.2 Case study: Detection of cobalt and chromium ions in patients with metal-on-metal implants *61*
3.2.3 FTIR 63
3.2.3.1 Case study: FTIR study of silanol groups in silica, slica-alumina, and zeolites *64*
3.2.3.2 Case study: Quantification of bridging and non-bridging Si—O as a function of SiO_2 % by FTIR *65*
3.2.4 X-Ray photoelectron spectroscopy (XPS)—surface chemistry 1 66
3.2.4.1 Case study: XPS study on Si—O—Si bridging energy variation *67*
3.2.5 Secondary ion mass spectrometry (SIMS)—surface chemistry 2 68
3.2.5.1 Case study: Investigation of the surface chemistry of a bioglass-polymer hybrid composite *69*
3.3 Physical Property Analysis 72
3.3.1 X-Ray diffraction 72
3.3.1.1 Case study: Characterization of a calcium hydroxyapatite reference material *74*
3.3.1.2 Case study: Characterization of amorphous and crystalline materials *75*
3.3.2 Nanoporosity characterization 75
3.3.2.1 Case study *80*
3.4 Microscopy 82
3.4.1 SEM, BEM and EDX 82
3.4.1.1 Case study: Drug-loaded sol-gel glass particles *83*
3.4.2 TEM 83
3.4.2.1 Case study: Drug-loaded mesoporous silica *84*
3.5 Technology for Evaluation of API and inorganic CRT Carrying System 84
3.5.1 Evaluation of API and inorganic CRT carrying system 84
3.5.2 Evaluation of API purity 85
3.5.2.1 Case study: Determination of genotoxic impurities in an API batch *86*

Inorganic Controlled Release Technology
http://dx.doi.org/10.1016/B978-0-08-099991-3.00003-X

3.5.3 Toxicity evaluation 86
3.5.3.1 An active material 88
3.5.3.2 A new excipient 89
3.5.3.3 A known excipient 89
References 90
Further Reading 91

3.1 INTRODUCTION

A comprehensive materials characterization is critical in supporting materials selection, research, product development, manufacturing and associated problem solving relevant to materials and product performance. A thorough understanding and employing of the most advanced materials analysis techniques are important too, for R&D and manufacturing in general. In-depth knowledge of critical material characteristics will help researchers to reach solutions for issues under investigation speedily. It is important therefore to understand firstly the basics and principles of each chosen technology to be used and, secondly, to be able to interpret conclusively the results produced by each characterization technology employed. This will assist scientists, technologists, engineers, QA/QC professionals, and relevant managers to make informed and correct decisions on design, selection, formulation and processing of materials.

Techniques for materials characterization need to be as advanced and powerful too. Appropriate use of such technologies can bring new insights into working principles and science of inorganic materials that are closely associated with drug release and the relevant evaluation. This includes examination of the bulk and surface properties.

There are many books that have extensively documented a range of methodologies used for materials characterization. It is not the intention to repeat all of that technology here, but to discuss some of the most important techniques *mainly through physical means* to achieve full understanding of materials characterization. We will introduce basic working principles and application examples that are relevant to the characterization of inorganic materials. For example:

- X-ray diffraction (XRD) and X-ray fluorescence (XRF), which can be employed in inorganic compound (glass and ceramic) testing through to bulk sample analysis.
- Scanning electron microscopy (SEM), which can provide a high-resolution overview of the surface.

- X-ray photoelectron spectroscopy (XPS), which can be employed in surface analysis and provides quantified elemental and oxidation-state information.
- Time-of-flight secondary ion mass spectrometry (ToF-SIMS), which can reveal detailed molecular information from the outer nanometres of the surface.
- Fourier transform infrared (FTIR), which can be used for a range of chemical analysis and can be used to detect bridging and non-bridging oxygen in glass.

We will group the relevant analytical methods under the following topics:

- chemical analysis,
- physical characterization covering both bulk and surface analysis,
- microscopy, and
- API (active pharmaceutical ingredient) evaluation.

3.2 CHEMICAL ANALYSIS

3.2.1 X-Ray Fluorescence

X-ray fluorescence (XRF) is the technique of choice for identifying, and in most cases quantifying, the inorganic components in a range of materials, such as a phosphate or silicate glass. It is the only technique for elemental analysis that allows the direct measurement of a sample without vaporization and prior digestion.

Working principle: X-rays form part of the electromagnetic spectrum and are expressed in terms of either their energy (kiloelectron volts—keV) or wavelength (nanometres—nm). When a high-energy incident (primary) X-ray collides with an atom, it disturbs the ground state conformation of the constituent nucleus and electrons and the atom is elevated to an excited state. Orbiting electrons are organized into shells: each shell is made up of electrons with the same energy. Following the interaction of an atom with the incident X-ray, an electron can be ejected from a low energy level and a 'hole' is created. An electron from a higher energy level can then fall into this space and a photon is released which is equal in energy to the energy difference between the two atomic orbitals involved in the electron transition. Therefore the obtained XRF spectrum reveals all the possible electron transitions taking place when a sample is bombarded with incoming X-ray radiation. Due to the unique atomic structure of individual elements, the number of protons, neutrons, and electrons present, the resultant XRF spectrum is characteristic of the specific elements present in a sample.

3.2.1.1 *Case Study: Contamination Investigation*

Table 3.1 shows typical results obtained from an XRF analysis. The material in question is a P-glass (phosphate glass); the theoretical composition is listed in Table 3.1a and the experimentally obtained XRF results are shown in Table 3.1b. The control P-glass sample contains no SrO while samples 1 and 2 contain 5 and 10 mol% of SrO, with the MgO content reducing from 30% down to 25% and 20%, respectively. In the original formulation no SiO_2 was present, but as can be seen from the XRF results, the glass products now contain ~2 mol% SiO_2. This SiO_2 contamination is likely to have been introduced during the glass melting process; the high temperature (1200 °C) required to form the homogeneous phosphate glass melt dissolved or degraded a portion of the crucible surface, which was made of highly pure SiO_2.

Limitation: Detection limits for XRF are generally restricted to 0.01% or 0.02% and therefore cannot detect the presence of components present in trace amounts, ppm and below. This is very important for the analysis of materials which may contain very low quantities of, for example, heavy metal ions such as cadmium and lead where even ppm amounts can cause significant toxicological and compliance issues. In instances where this high level of chemical analysis resolution is required, techniques such as inductively coupled plasma optical emission spectroscopy would be more suitable.

3.2.2 Inductively Coupled Plasma Mass Spectrometry

Inductively coupled plasma mass spectrometry (ICP-MS) has received significant attention more recently and has been widely used for the analysis of metallic, inorganic as well as organic and biological samples. When

Table 3.1 Phosphate glass composition evaluation

Phosphate glass	P_2O_5	CaO	Na_2O	MgO	SrO	SiO_2
(a) Theoretical composition (mol%)						
Control	40	25	5	30	0	0
Sample 01	40	25	5	25	5	0
Sample 02	40	25	5	20	10	0
Phosphate glass	**P_2O_5**	**CaO**	**Na_2O**	**MgO**	**SrO**	**SiO_2**
(b) XRF experimental results (mol%)						
Control	40.2	23.7	5.1	28.9	0.0	2.1
Sample 01	39.7	23.9	5.0	24.8	4.2	2.4
Sample 02	40.5	23.5	5.0	19.4	8.9	2.7

compared with other techniques, the greater advantage of ICP-MS is its high sensitivity with extremely low limits of detection, in the range of nmol L^{-1} for several elements, and multielemental capability.

Working principle: ICP-MS is a technique that combines two technologies into one:

- A high-temperature ICP source, which is in a range between 6000 and 10,000 K. The high-temperature ICP source converts the atoms of the elements in the sample to ions.

These ions are then separated and detected by

- a mass spectrometer. The ions formed by the ICP discharge are typically positive ions, M^+ or M^{2+}.

3.2.2.1 Case Study: Controlled Release of Strontium from P-glass

Due to a globally increasing ageing population, surgical intervention using orthopaedic bone implants in patients who are generally older and more osteoporotic is increasing. These patients typically will have a slower osteogenesis capacity. To aid the bone healing processes at the disease or defect sites, the addition of stimulus ions such as strontium (Sr) is being well researched and is thought to facilitate the recovery of diseased bone. ICP-MS is a good tool by which to study the kinetics of the active ion release. Take the formulation Sample 01 in Table 3.1 as an example. The P-glass was placed in aqueous media for a pre-determined length of time and the weight loss of the glass powder was determined. Vials with glass powder and medium were set to 37 °C with a shaking speed of 100 rpm. Lucideon's ICP-MS analytical facility was used to monitor the kinetics of ion release, which is shown in Figure 3.1, a plot of strontium ion concentration as a function of time. Analysis of the release profile suggests that the diffusion kinetics is very close to Case III diffusion (refer to Section 2.4 in Chapter 2), i.e. the index n is close to 1 based on Equations (2.13) and (2.14).

3.2.2.2 Case Study: Detection of Cobalt and Chromium Ions in Patients with Metal-on-Metal Implants

The potential for release of metal debris, in particular cobalt and chromium, from metal-on-metal hip implants into the bloodstream has become a serious concern in recent times. It is understandable that the liberated metal particles are likely to be present in only very low concentrations, but the long-term medical complications arising from such wear are as yet unknown. The question is: how to quantify such very low concentrations of the metal ions? The only possible way is to quantitatively analyze the blood for the presence of

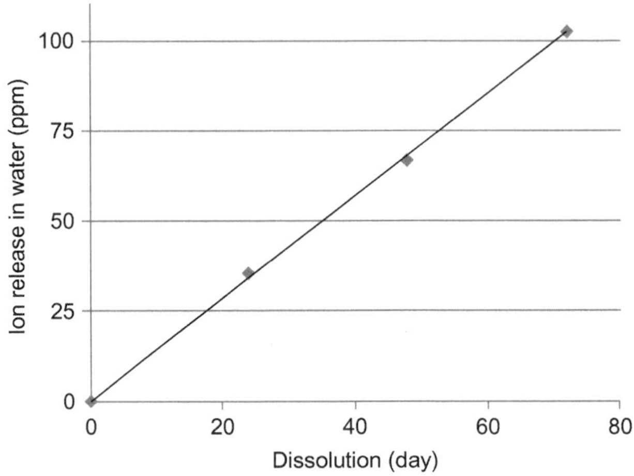

Figure 3.1 Example of controlled strontium release from a P-glass.

metal debris. In this respect, no other technique can match the performance of ICP-MS for characterization and quantification of very low concentrations of specific elements. Clarke[1] used the technology successfully to quantify the concentration of cobalt and chromium ions for a range of patient groups who had metal-on-metal implants. Blood samples were taken from patients, centrifuged at 3000 rpm for 10 min, and then the separated plasma was frozen at −80 °C, ready for analysis. A Perkin-Elmer SCIEX Elan 6100 DRC plus was employed, yielding the following results:

- median cobalt level:
  - 22 nmol L^{-1} for normal hip implants with a head diameter of 28 mm. This cobalt concentration is 4.4 times the acceptable upper limit of 5 nmol L^{-1};
  - 38 nmol L^{-1} for the hip implants that were done by resurfacing. This concentration is 7.6 times the acceptable upper limit;
- median chromium level:
  - 19 nmol L^{-1} for normal hip implants of 28 mm diameter head. This concentration is 3.8 times the average top level of healthy people who do not have such implantation;
  - 53 nmol L^{-1} for resurfacing surgery, which is 10.6 times the normal level.

These results showed that following the introduction of metal-on-metal hip implants, both the cobalt and chromium concentrations were well above the upper bloodstream limit of 5 nmol L^{-1}. This is just one study highlighting some of the long-standing safety concerns associated with metal-on-metal hip implants.

This is a good example of the use of ICP-MS to inform the debate on a particularly complicated medical health research topic.

Limitation: Detection capabilities may vary with the sample matrix, which may affect the degree of ionization that will occur in the plasma, or allow the formation of species that may interfere with the analyte determination. Elements that prefer to form negative ions, such as chlorine, iodine, and fluorine, are more difficult to determine *via* ICP-MS.

3.2.3 FTIR

Fourier transform infrared (FTIR) spectroscopy is a common technique used in a wide variety of application areas, not least for materials analysis. An infrared spectrum represents a fingerprint of a sample with absorption peaks that correspond to the frequencies of vibrations between the bonds of the atoms making up the material. Because each different material is a unique combination of atoms, no two compounds produce the exact same infrared spectrum. Therefore, infrared spectroscopy can result in a positive identification (qualitative analysis) of every different kind of material. In addition, the size of the peaks in the spectrum is a direct indication of the number of a specific bond present. With the aid of modern software algorithms, infrared is an excellent tool for quantitative analysis.

FTIR is particularly useful for the identification of organic molecular groups and compounds due to the range of functional groups, side chains, and cross-links that are inherent with these types of materials, all of which will have characteristic vibrational frequencies in the infrared range. Even so, we will try to contextualize the use of FTIR spectroscopy for the purpose of materials research and development for inorganic CRT applications.

The primary advantages of using FTIR, in general, include speed and sensitivity; in principle all of the frequencies are measured simultaneously, and this can be achieved in a few seconds. The modern detection systems employed are much more sensitive and the optical throughput is much higher, leading to much lower noise levels.

Working principle: FTIR relies on the fact that most molecules absorb light in the infrared region of the electromagnetic spectrum. This absorption corresponds specifically to the bonds present in the molecule, with frequencies measured as wave numbers, typically over the range 4000-600 cm^{-1}. The background emission spectrum of the IR source is first recorded, followed by the emission spectrum of the IR source with the sample in place. The ratio of the sample spectrum to the background spectrum is directly related to the sample's absorption spectrum. The absorption spectrum resulting from the specific bond vibration frequencies indicates the presence of various chemical bonds and functional groups in the sample.

3.2.3.1 Case Study: FTIR Study of Silanol Groups in Silica, Slica-Alumina, and Zeolites

Janin and co-workers employed FTIR to study the presence of silanol groups in dealuminated HY zeolites. Hydroxyl groups exist in different forms in HY zeolites including as Al—OH groups, two types of bridging Si—OH—Al groups and Si—OH groups.[2] A detailed FTIR study confirmed the presence of these three different types of silanol groups. In general,

(i) a first band at 3747-3749 cm^{-1}, which is assigned to SiOH groups attached to amorphous silica-alumina debris. They are sensitive to pyridine adsorption and desorption at 423 K (refer to Figure 3.2);

(ii) a second band at 3744-3746 cm^{-1} corresponding to silanol groups attached to extra framework silicon-rich debris. They are not sensitive to pyridine;

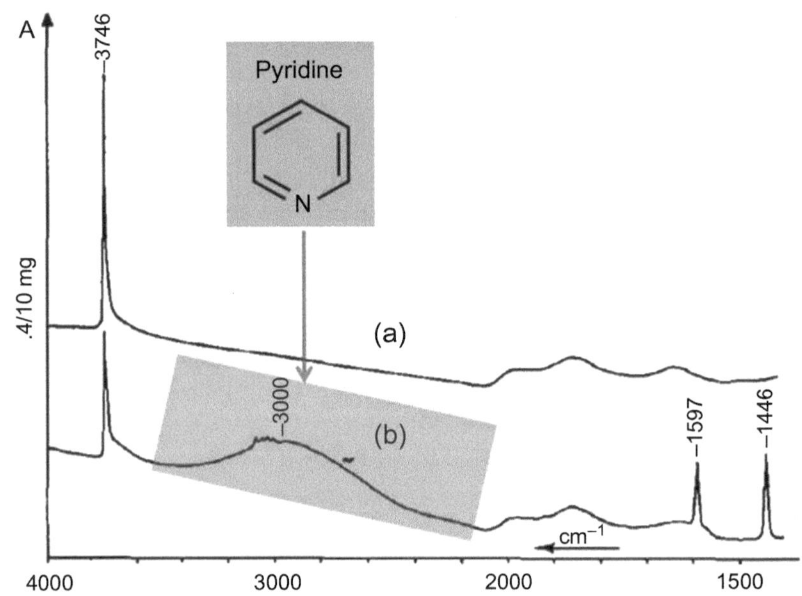

Figure 3.2 Pyridine adsorption on SiO_2: (a) control—before pyridine adsorption and (b) after pyridine absorption at room temperature. *(Reprinted from Ref. 2 with permission from Elsevier).*

(iii) a third component at 3738 cm^{-1} assigned to terminal framework silanol groups. They are not sensitive to pyridine either.

The silanol bands present in modified zeolites provide information on the nature of the extra framework debris and on framework structural defects.

Figure 3.2 shows an example of an FTIR spectrum for amorphous silica. The spectrum exhibits a strong Si—OH band at 3746 cm^{-1}. Further Si—OH groups were revealed with the aid of pyridine adsorption onto the silica (Figure 3.2b). They are characterized by a broad band at 2950 cm^{-1}. This is because pyridine leads to strongly hydrogen-bonded species with Si—OH groups that are more clearly seen in FTIR spectra. Although detailed structural information on the molecular level is not well known, the variation in the form of Si—OH groups present should be seen in the spectra (also refer to silanol modelling in Section 2.3.4 in Chapter 2). Figure 3.2 also shows the absorptions at 1597 and 1446 cm^{-1}, respectively, which show the adsorption of pyridine.

3.2.3.2 Case Study: Quantification of Bridging and Non-bridging Si—O as a Function of SiO_2 % by FTIR

Serra[3] and co-workers reported on how the SiO_2 content of bioactive glasses affects the ratio of non-bridging Si—O based on the SiO stretching vibration in the FTIR spectrum. This is shown in Figure 3.3, which confirms that increasing SiO_2 content decreases the ratio of non-bridging SiO—NB over bridging SiO—B.

In general, it is expected that FTIR technology can detect bridging stretching vibrations of X-O-X and non-bridging vibrations with variation of the bioglass formulation, where X is the major glass network former. This is an area that has only a limited number of previous studies. Three types of glass, named after their network former families (refer to Table 1.2, Chapter 1), silica, phosphate, and boron glasses, are of particular interest in this respect. In addition to these three network former oxides, it is also worth noting that some intermediate oxides may act as glass network formers depending on which other glass former and network modifying ions are present in the melt. For example, arsenic and antimony form glasses in the presence of intermediates of other oxides if cooled quickly enough. Therefore, it is necessary to carry out more research to gain insights into the bonding energy variations influenced by different cations on the network formers (Figure 2.14 in Chapter 2) and to quantify these variations for new inorganic CRT technologies in order to make best use of the technology for a range of medical applications.

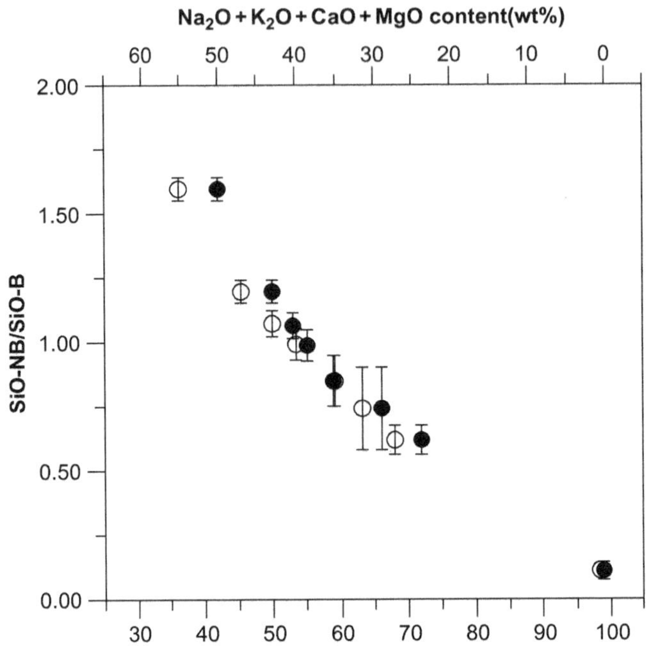

Figure 3.3 Effect of bioglass formulation on the absorption intensity ratio of non-bridging SiO—NB over bridging SiO—B. *(Reprinted from Ref. 3 with permission from Elsevier).*

Limitation: The main limitation of FTIR is that it is difficult to quantify the amounts of certain elements when they are present in low concentrations; its sensitivity is in the range of 0.1-1%.

3.2.4 X-Ray Photoelectron Spectroscopy (XPS)—Surface Chemistry 1

Photoelectron spectroscopy utilizes photo-ionization and analysis of the kinetic energy distribution of the emitted photoelectrons to study the composition and electronic state of the surface region of a sample. For each and every element, there will be a characteristic binding energy associated with each core atomic orbital, and as a result, each element will give rise to a characteristic set of peaks in the photoelectron spectrum at kinetic energies determined by the photon energy and the respective binding energies.

Working principles: XPS spectra are obtained by irradiating a material with a beam of X-rays while simultaneously measuring the kinetic energy and number of electrons that escape from the top 0 to 10 nm of the material being analyzed. Each element produces a characteristic set of XPS peaks at characteristic binding energies that directly identify each element that exists in or on the surface of the material being analyzed. These characteristic spectral peaks correspond to the configuration of the electrons within the atoms. The number of detected electrons in each of the characteristic peaks is directly related to the amount of element within the XPS sampling volume. XPS works based on a single photon in and electron out principle. The total photo energy can be expressed by

$$E_{\text{photo}} = E_{\text{kinetic}} + E_{\text{binding}} \tag{3.1}$$

where E_{photon} is the energy of the X-ray photons being used, E_{kinetic} is the kinetic energy of the electron, and E_{binding} is the binding energy of the electron.

Rearranging Equation (3.1) gives

$$E_{\text{binding}} = E_{\text{photo}} - E_{\text{kinetic}} \tag{3.2}$$

which is used to calculate the binding energy. In practice, Equation (3.2) must include another instrument correction factor $E_{\text{instrument}}$. So the final equation for the bonding energy is given by

$$E_{\text{binding}} = E_{\text{photo}} - (E_{\text{kinetic}} + E_{\text{instrument}}) \tag{3.3}$$

Because the photo energy of an X-ray with a particular wavelength is known, the kinetic energy can be measured, and the instrument correction factor is a constant, the bonding energy can be calculated *via* Equation (3.3).

3.2.4.1 Case Study: XPS Study on Si—O—Si Bridging Energy Variation

Based on the same formulation and FTIR results shown in Figure 3.3, Serra and co-workers were successful using the XPS technology to work out the bonding energy variation as a function of SiO_2 content. This is shown in Figure 3.4.

Raskar[4] and co-workers studied structural organization of sodium borophosphate glasses with composition $(NaPO_3)_{1-x} - (B_2O_3)_x$ $(0.0 \leq x \leq 0.3)$. O^-1s XPS data provide a quantitative distinction between B—O—B, B—O—P, and P—O—P linkages as well as non-bridging oxygen atoms within the compositional region $0 \leq x \leq 0.20$. The entire boron content is present in the form of anionic BO_4^- units, resulting in the repolymerization of an equivalent fraction of the phosphate units. XPS reveals strong interactions between the two network formers, boron oxide and phosphorus oxide, resulting in the dominant formation of B—O—P linkages. The enhanced degree of network polymerization correlates with a significant increase of the glass transition temperature as a function of boron content.

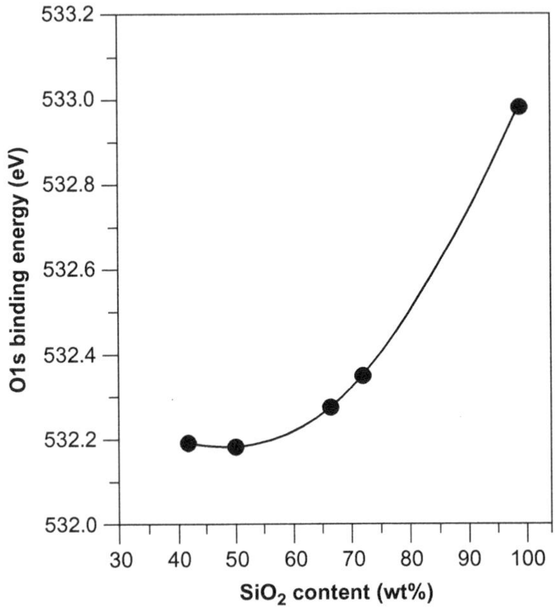

Figure 3.4 Effect of SiO_2 content on binding energy Si—O—Si. *(Reprinted from Ref. 3 with permission from Elsevier).*

In general, the capacity to measure bonding energy to great resolution will make the XPS technology a useful tool in obtaining compositional, structural, and bonding information to details with high resolutions not achievable by other analytical methods.

Limitation: It is difficult to quantify the presence of species in low concentrations with XPS. Its sensitivity is in the range of 0.1 (atomic) %. Additionally, XPS has limited sample depth penetration—only the top 10 nm can be sampled.

3.2.5 Secondary Ion Mass Spectrometry (SIMS)—Surface Chemistry 2

SIMS is an analytical technology that has much higher sensitivity in comparison with XPS—down to ppm-ppb levels of resolution—and can be used to detect nearly all of the elements of the periodic table, from hydrogen to uranium. Typically, the volume of material sputtered is small. Using a high-energy and highly focussed beam of primary ions (dynamic SIMS) a volume of ~100-1000 μm^3 is analyzed. Alternatively when using low-energy, low-density primary ion beams (static SIMS), the volume of material sputtered is exceedingly low with only the surface monolayers being sampled. So, highly accurate analysis is achievable at a point close to the area of interest. In another variation on the SIMS technique, a pulsed ion beam can be used to remove molecules from the very outermost surface of the sample. The secondary ions generated are then accelerated into a 'flight tube' and their mass is determined by measuring the exact time at which they reach the detector. This specific SIMS technique is known as time-of-flight secondary ion mass spectrometry—ToF-SIMS.

Working principle: Figure 3.5 schematically shows how primary beam ions interact with a sample to remove material from the surface. The SIMS technology is capable of sputtering the sample on the very top of the surface, from 1 to 2 nm. It can also sample the material layer by layer up to 100 μm. The ions generated during this process are then detected by mass spectrometry. Both positively and negatively charged ions can be analyzed routinely, which is an advantage over the use of ICP-MS analytical technology (refer to Section 3.2.2).

3.2.5.1 Case Study: Investigation of the Surface Chemistry of a Bioglass-Polymer Hybrid Composite

Figure 3.6 shows a typical mass spectrum resulting from a ToF-SIMS analysis and is representative of a bioactive glass embedded within polycaprolactone. The bioactive glass material was made by fusion technology, the composition

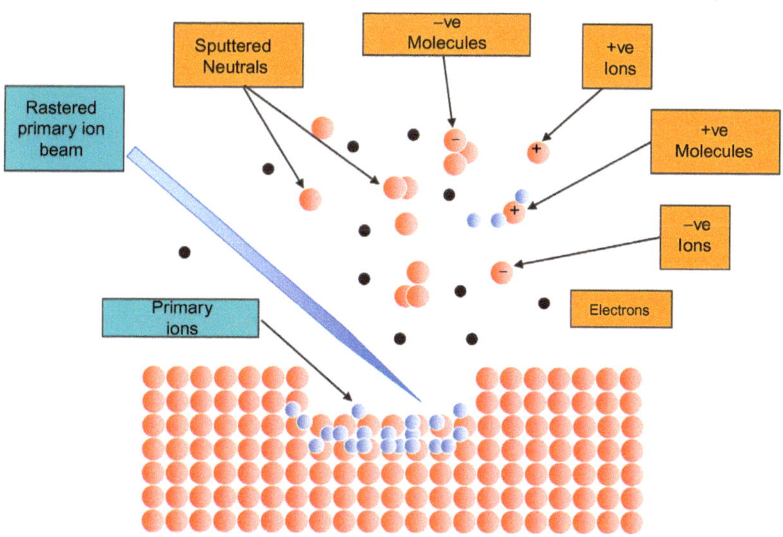

Figure 3.5 Sputtering the sample with an ion beam to remove material from the top 1-2 nm or layer by layer up to 100 μm. *(Courtesy of Lucideon Surface Science Department).*

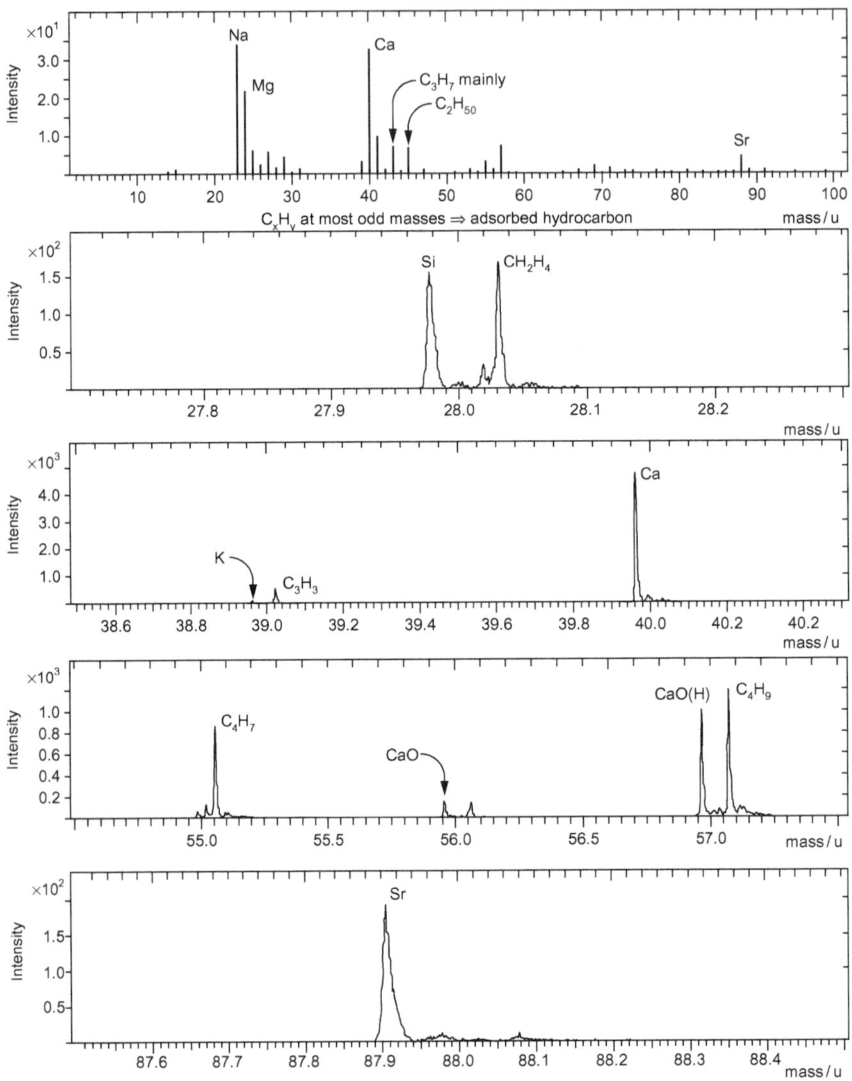

Figure 3.6 ToF-SIMS analysis of bioactive glass P_2O_5—CaO—MgO—SrO—SiO_2 embedded in polycaprolactone: (a) positively charged ions and

Continued

of which consists of P_2O_5—CaO—MgO—SrO—SiO_2. The area of the analysis was about 0.2 mm × 0.2 mm. The mass spectra are results from 'soft'-SIMS, that is, the primary beam only removed the very top surface of the sample, about 1 nm. Single ions (positive or negative), individual isotopes, and molecular clusters were generated and detected at the time they formed with high mass resolution (>10,000-accurate mass) and high sensitivity (ppb) for

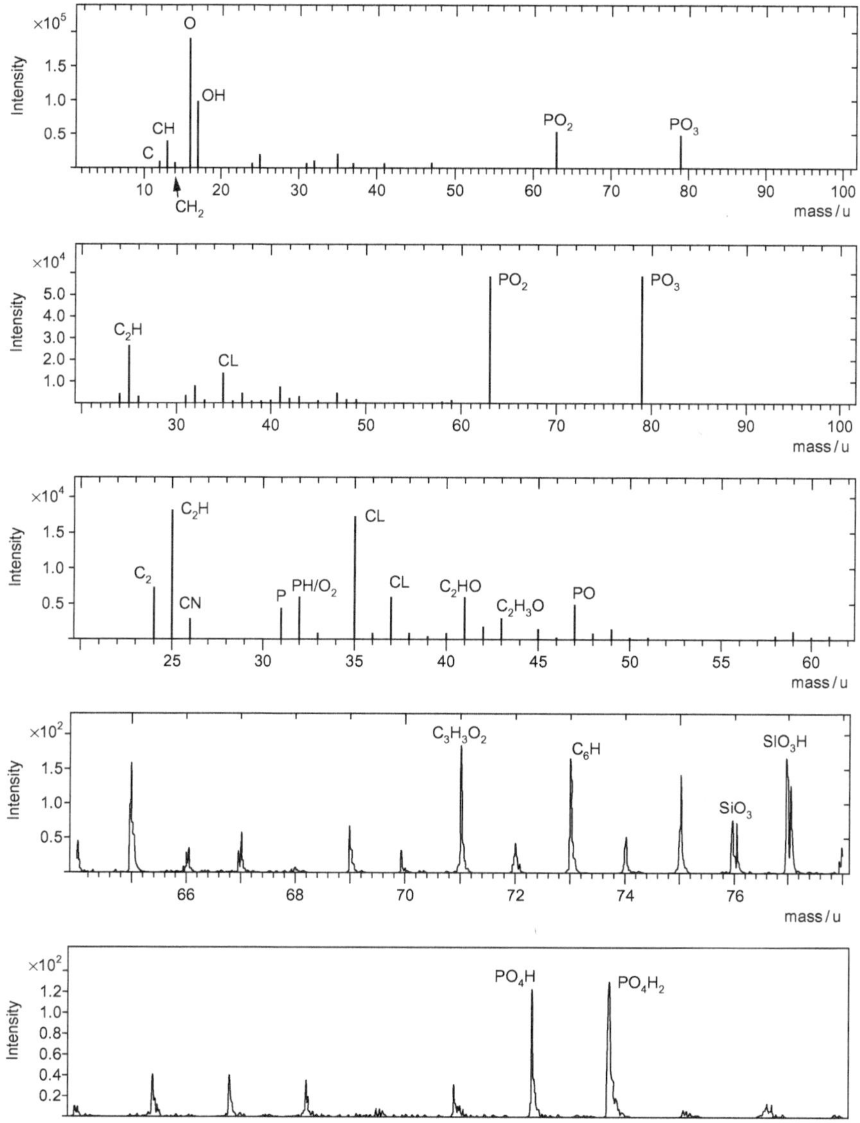

Figure 3.6, cont'd (b) negatively charged ions; CL = caprolactone. *(Courtesy of Lucideon Surface Science Department, special thanks to Dr Alan Paul who helped to assign the ions listed).*

trace elements or compounds. The ToF-SIMS analysis has collected both a positive and a negative spectrum to give the complete picture over 1000 mass/u. Figure 3.6 shows only the mass spectrum at low mass/charge ratio, including

- single ions: Na, K, Mg, Ca, Si, Sr, O;
- identified single groups: inorganic includes CaO, MgOH, CaO(H), SiO(H), SrO(H), CaH, PO_2, PO_3, SiO_3, SiO_3H, PO_4H, PO_4H_2;
- organic segments associated with polycaprolactone.

We will not go through detailed quantitative analysis here. The high sensitivity and resolution of the ToF-SIMS technique provides good chemical information on the surface of the glass. It is expected that the results achieved here for the surface characteristics will be different from those of the bulk. For instance, it is not expected that the following ion groups associated with H could be detected in bulk, such as MgOH, CaO(H), SiO(H), SrO (H) and CaH. The detection of these ions might indicate the presence of —OH groups and/or water that may not be completely removed under long, high vacuum conditions.

Limitations: SIMS is a destructive technique and no chemical bonding information can be obtained in comparison with XPS. Generally it is difficult to produce quantitative analyses.

3.3 PHYSICAL PROPERTY ANALYSIS

3.3.1 X-Ray Diffraction

X-ray diffraction (XRD) is often used to discern the crystal lattice structure of the different solid state forms of a material. The discovery of X-rays in 1895 enabled scientists to probe crystalline structure at the atomic level. XRD has found extensive use in two main areas: for the fingerprint characterization of crystalline materials and the determination of their structure. Each crystalline solid has its unique characteristic X-ray powder pattern that may be used as a 'fingerprint', analogous to that for FTIR for a materials chemical identification. Once the material has been identified, X-ray crystallography may be used to determine its structure, that is, how the atoms pack together in the crystalline state and what the interatomic distance is. Today about 50,000 inorganic and 25,000 organic single components, crystalline phases, and diffraction patterns have been collected and stored on magnetic or optical media as standards (http://www.ccdc.cam.ac.uk/pages/Home.aspx—the Cambridge Crystallographic Data Centre is one of the most important online repositories for crystal structures).

Working principles: A solid matter can be described as amorphous and/or crystalline or a combination of the two, which has been discussed in Section 2.1. For a perfect crystal, the atoms are arranged in a regular three-dimensional pattern, the basis of which is a smallest volume element known as the unit cell. Infinite repetition of the unit cell in all three dimensions results in a perfect crystal. Crystals are regular arrays of atoms, while X-rays can be considered waves of electromagnetic radiation. Atoms scatter X-ray waves, primarily through the atoms' electrons. An X-ray striking an electron produces secondary spherical waves emanating from the electron. This phenomenon is known as elastic scattering, and the electron is known as the scatterer. A regular array of scatterers produces a regular array of spherical waves. Although these waves cancel one another out in most directions through destructive interference, they add constructively in a few specific directions, as determined by Bragg's law:

$$2d\sin\theta = n\lambda \qquad (3.4)$$

This is illustrated in Figure 3.7, which is based on the perfect crystal structure of Figure 2.1a. Here we highlight an X-ray beam of incident wavelength λ on a pair of parallel planes plane 1 and plane 2 at an angle of θ, which is separated by an interplanar spacing d of the crystal.

Bragg's law states that a reflected beam of maximum intensity will be 'seen' only if the X-ray waves represented by $1'$ and $2'$ are in phase. The difference in path length between 1 to $1'$ and 2 to $2'$ must be an integral number, which is the n in Equation (3.4). Consideration of two important points will aid the understanding here:

- the angle of incidence = angle of scattering;
- the path length difference = $n\lambda$, where n must be an integer.

Therefore, with the aid of Figure 3.7, we see that possible 2θ values where we can have constructive reflections are determined by the unit cell

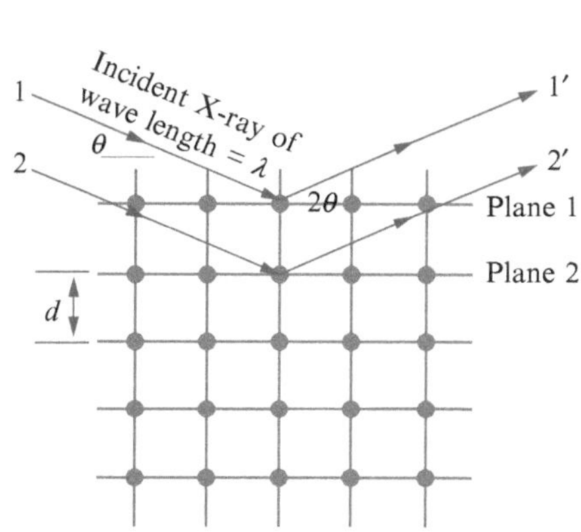

Figure 3.7 Schematic illustration of X-ray interaction with atoms in a two-dimensional crystal structure.

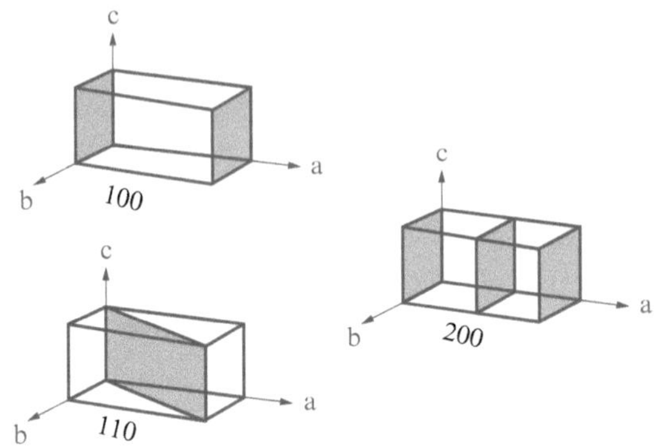

Figure 3.8 Schematic illustration of unit cells with assigned Miller indices (*hkl*).

dimensions. However, the intensities of the reflections are determined by the distribution of the electrons in the unit cell. The highest electron density is found around atoms and therefore the intensities depend upon what kinds of atoms are present in the unit cell and where exactly they are located.

X-rays are used to produce the diffraction pattern because their wavelength λ is typically of the same order of magnitude (1-100 Å) as the spacing d between planes in the crystal. So the XRD technology allows visualization of the crystal structure at an atomic level.

Let us now move on from 2D to a simplified 3D crystal unit cell model as shown in Figure 3.8. This model shows X-ray reflections from a series of parallel planes inside the crystal. The orientation and interplanar spacings of these planes are defined by the three integers h, k and l, called Miller indices. For example, the (1 0 0) plane cuts axis a but is parallel to the b and c axes; the (2 0 0) plane cuts the a axis in half but is parallel to the b and c axes; the (1 1 0) plane cuts axes a and b but is parallel to c. In general, a zero index indicates that the planes are parallel to the corresponding axis.

3.3.1.1 Case Study: Characterization of a Calcium Hydroxyapatite Reference Material[5]

Calcium hydroxyapatite (HA), $Ca_{10}(PO_4)_6(OH)_2$, is an important inorganic material in biology and chemistry. Biological apatites, which are the inorganic constituents of bone, tooth enamel, and dentin, are typically very variable in their composition and morphology, and typically contain different impurities. In general, these impure biological apatites are designated as calcium deficient or non-stoichiometric apatites. Synthetic HAs are frequently used as reference materials in biomineralization and biomaterial studies. The composition, physicochemical properties, crystal size, and morphology of synthetic apatites are extremely sensitive to preparation conditions. For control and reference purposes, it is important to have available pure and stoichiometric HA, or nearly stoichiometric HA, characterized in detail with respect to its chemical composition and numerous other important properties. To meet this need, a large amount of highly pure, homogeneous, and highly crystalline HA needs to be synthesized and evaluated by XRD technology.

The XRD pattern of hydroxyapatite is shown in Figure 2.2 and is a spectrum resulting from a standard HA. The results are collected in Table 3.2, which includes 2θ-values and relative intensities (I_{rel}) observed from the XRD pattern of HA, d-values determined from 2θ-values ($d_{2\theta}$), d-values calculated from unit cell parameters (d_{calc}), and the corresponding indices (hkl). From this analysis, the hexagonal HA crystal cell unit was determined; the average values are $a=b=0.94235$ nm and $c=0.68852$ nm.

3.3.1.2 Case Study: Characterization of Amorphous and crystalline Materials

Figure 2.1 represents two different atomic structures: one is a pseudoperfect crystal and one is for an amorphous material. A particular beneficial feature of XRD analysis is that it is a useful tool to confirm if a material or a mixture of several materials is/are 100% amorphous or not. If not, XRD can be used to quantify the crystallinity of the material or the mixture of materials. For inorganic CRT applications, XRD has been used to detect whether a mesoporous silica or phosphate glass contains any crystal structure prior to drug loading. After drug loading, XRD can also be used to confirm if the drug is dispersed and distributed into the nanopores to a pseudomolecular state or not because most drugs will tend to their most energetically favourable crystalline state. Figure 3.9 is a good example demonstrating the presence of crystalline drug within an amorphous inorganic CRT delivery system.

Limitation: Limitations of XRD are that it cannot be used to analyze non-solid materials or quantify the relative proportions of crystalline materials if they are present in low concentrations.

3.3.2 Nanoporosity Characterization

Surface area, pore size and distribution are very important parameters in determining the quality of an inorganic CRT system, particularly when the nanopore size is a factor in designing the properties of a formulation, such as is the case with mesoporous silica. The measurement is normally

Table 3.2 2θ-values and relative intensities (I_{rel}) observed from the XRD pattern of HA, d-values determined from 2θ-values ($d_{2\theta}$), d-values calculated from unit cell parameters (d_{calc}) and corresponding indices (hkl)

2θ	$d_{2\theta}$ (nm)	d_{calc} (nm)	I_{rel} (nm)	(hkl)
10.85	0.815	0.816	8	100
16.87	0.525	0.526	3	101
18.84	0.471	0.471	2	110
21.75	0.408	0.408	6	200
22.84	0.389	0.389	6	111
25.35	0.351	0.351	2	201
25.86	0.344	0.344	36	002
28.11	0.317	0.317	8	102
28.92	0.308	0.308	16	210
31.77	0.281	0.281	100	211
32.18	0.278	0.278	47	112
32.90	0.272	0.272	65	300
34.04	0.263	0.263	22	202
35.44	0.253	0.253	5	301
39.18	0.2297	0.2297	6	212
39.793[a]	0.2263	0.2263	22	310
40.43	0.2229	0.2229	1	221
41.98	0.2150	0.2150	6	311
42.30	0.2135	0.2134	1	302
43.84	0.2063	0.2063	4	113
44.36	0.2040	0.2040	1	400
45.29	0.2000	0.2000	4	203
46.683[a]	0.1944	0.1944	28	222
48.068[a]	0.1891	0.1891	12	312
48.58	0.1872	0.1872	3	320
49.458[a]	0.1841	0.1841	30	213
50.474	0.1807	0.1807	15	321
51.254[a]	0.1781	0.1781	11	410
52.061[a]	0.1755	0.1755	11	402
53.167[a]	0.1721	0.1721	14	004
54.43	0.1684	0.1684	1	104
55.85	0.1645	0.1645	6	322
57.11	0.1611	0.1611	4	313
58.03	0.1588	0.1588	2	501
58.28	0.1582	0.1582	2	412
58.74	0.1570	0.1570	1	330
59.93	0.1542	0.1542	4	420

[a] *Note*: 2θ-values have an expanded uncertainty (U) of $\pm 0.004°$ 2θ ($n=4$).

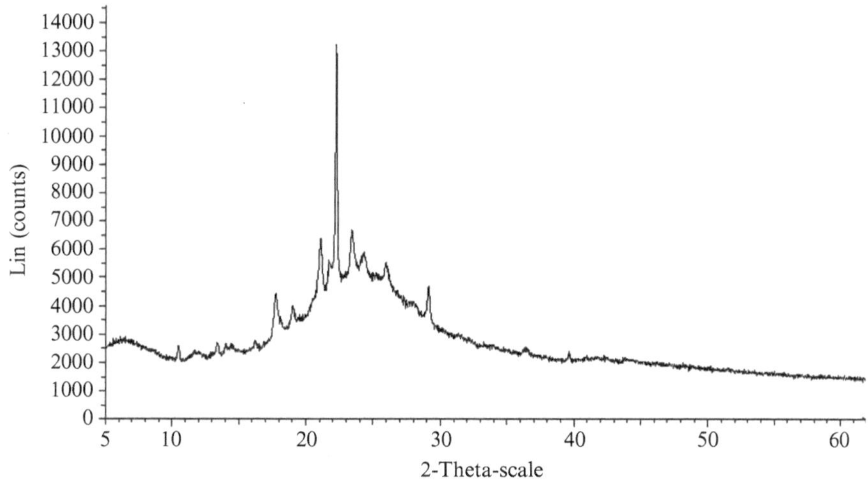

Figure 3.9 XRD diffractogram showing the presence of crystalline material (sharp peaks) within an amorphous inorganic CRT carrier system (broad halo at ~20-30 2θ). *(Spectrum courtesy of Lucideon X-ray Department).*

carried out by using nitrogen adsorption/desorption isotherms at the temperature and pressure of liquid nitrogen. An isotherm for a mesoporous material is shown in Figure 3.10. The large uptake of nitrogen at low P_e/P_o indicates the first-stage filling of the nanopores. The linear portion of the curve may represent multilayer adsorption of nitrogen on the surface of the nanopores. The later uptake of nitrogen shows the sample approaching full saturation of the pores. There are many theoretical and experimental interpretations on nanoporosity characterization; we will use a simplified mesoporous model to set a foundation to make readers aware of the most common parameters and how to handle the experimental data to gain useful information on the porosity characteristics of your material.

Working principles: Let us start with a simple model based on mesoporous silica assuming a regular cylindrical shape as shown in Figure 3.10.

Before nitrogen adsorption, the total surface area A_s and volume V_s of the nanocylinder can be expressed by

$$A_s = \sum_{i=1}^{n} A_i = \sum_{i}^{n} 2\pi l r_s = n 2\pi l r_s \tag{3.5}$$

$$V_s = \sum_{i=1}^{n} V_i = \sum_{i}^{n} \pi l r_s^2 = n\pi l r_s^2 \tag{3.6}$$

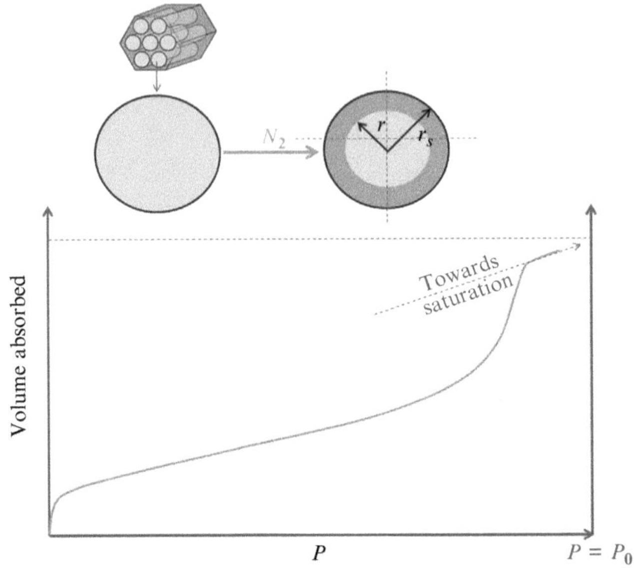

Figure 3.10 Modelling nitrogen adsorption onto a mesoporous material (P_0 is the saturation pressure).

where A_i is one nanocylinder surface area and V_i one nanocylinder volume, r_s is the radius of the cylinder, l is the length and n is the total number of nanocylinders.

Combination of Equations (3.5) and (3.6) gives

$$\frac{V_s}{A_s} = \frac{n\pi l r_s^2}{n2\pi l r_s} = \frac{r_s}{2} \tag{3.7}$$

$$r_s = 2\frac{V_s}{A_s}$$

$$d_s = 2r_s = 4\frac{V_s}{A_s} \tag{3.8}$$

Equation (3.8) is a general model often used to calculate the nanopore diameter. This is because both V_s and A_s can be experimentally obtained.

After adsorption of nitrogen, the thickness of the nitrogen layer adsorbed, that is, t_{N_2}, can be expressed by

$$t_{N_2} = r_s - r \tag{3.9}$$

The total inside surface area A_{N_2}, and volume V_{N_2} of nitrogen adsorbed into the nanopores can be expressed by

$$A_{N_2} = \sum_{i=0}^{n} 2\pi lr = n2\pi lr \tag{3.10}$$

$$V_{N_2} = \sum_{i=0}^{n} \left(\pi lr_s^2 - \pi lr^2\right) = n\left(\pi lr_s^2 - \pi lr^2\right) \tag{3.11}$$

Let us consider two special cases. *The first*, a complete monolayer adsorption, that is, $r = r_m$, Equation (3.9) becomes

$$t_m = r_s - r_m \tag{3.12}$$

where t_m is the thickness of N_2 monolayer adsorption. Because the dimension of the N_2 monolayer is so small, we can reasonably assume that $r_s \approx r_m$, so Equations (3.10) and (3.11) become

$$\lim_{r \to r_m} A_{N_2} = A_m \cong A_s = n2\pi lr_s \tag{3.13}$$

$$\lim_{r \to r_m} V_{N_2} = V_m = n\left(\pi lr_s^2 - \pi lr_m^2\right) = n\pi l(r_s + r_m)t_m \cong 2n\pi lr_s t_m = A_s t_m \tag{3.14}$$

where subscript 'm' stands for monolayer adsorbed N_2.

The second, full adsorption, that is, all pores are filled with N_2. So $r = 0$

$$\lim_{r \to 0} V_{N_2} = V_0 = V_s = n\pi lr_s^2 \tag{3.15}$$

So V_0 is the saturated volume.

BET model: The most famous model for dealing with gas adsorption and desorption data is the BET model.[6]

Let us first read the assumptions of BET theory to gain an insight into its fundamental principles:

1. Molecular adsorption:
   - **1.1** adsorptions occur only on well-defined sites of the sample surface, one per molecule;
   - **1.2** a molecule can act as a single adsorption site for a molecule of the upper layer;
   - **1.3** the uppermost molecule layer is in equilibrium with the gas phase.
2. Adsorption energy:
   - **2.1** the energy associated with adsorption of the first monolayer onto a solid surface is a constant for a given adsorbent, which is termed as E_s, that is, the heat of adsorption at the solid sample surface;
   - **2.2** the other layers above the first are assumed similar and can be represented as a condensed liquid phase, that is, liquid state. Hence, the heat of adsorption is equal to the heat of liquefaction, that is, E_l.

3. Saturation condition:
   3.1 the number of molecule layers tends to infinity, that is, equivalent to the sample being surrounded by a liquid phase, at the saturation pressure.

Based on the above assumptions, Brunauer, Emmett and Teller (BET) derived the following equation:

$$\frac{P_e}{V_{N_2}(P_0 - P_e)} = \frac{1}{V_m C} + \frac{C-1}{V_m C}\frac{P_e}{P_0} \tag{3.16}$$

where P_e and P_0 are the equilibrium and saturation pressures, V_m is the monolayer adsorbed gas volume, and C is the BET constant.

$$C \approx e^{(E_s - E_l/RT)} \tag{3.17}$$

By plotting Equation (3.16) on the left as a function of P_e/P_0, the slope and the intercept are given:

$$\text{Slope} = \frac{C-1}{V_m C} \tag{3.18}$$

$$\text{Intercept} = \frac{1}{V_m C} \tag{3.19}$$

With the aid of Equations (3.18) and (3.19), V_m is obtained.

In summary, porosity parameters of mesoporous materials can be measured by experimental approaches such as BET nitrogen adsorption and desorption techniques based on the above theories. The science and technology associated with nanopore structure characterization is not yet fully developed. Both theories and interpretation of testing results are still problematic to certain degrees in view of the fundamentals.[7] More work is needed to facilitate new nanoporous materials development programmes, particularly in the fastest growing market areas.

3.3.2.1 Case Study

Figure 3.11 is one experimental result obtained from mesoporous silica MCM-41 using the BET nitrogen adsorption method. Figure 3.11a is made by plotting adsorption volume V_{N_2} (μl/g) as a function of P_e/P_0. Figure 3.11a is a plot made using the left term of Equation (3.16) as a function of P_e/P_0.

(a) From the top right of Figure 3.11a, the saturated volume:
   - $V_0 = V_s = 0.5296$ μl/g.

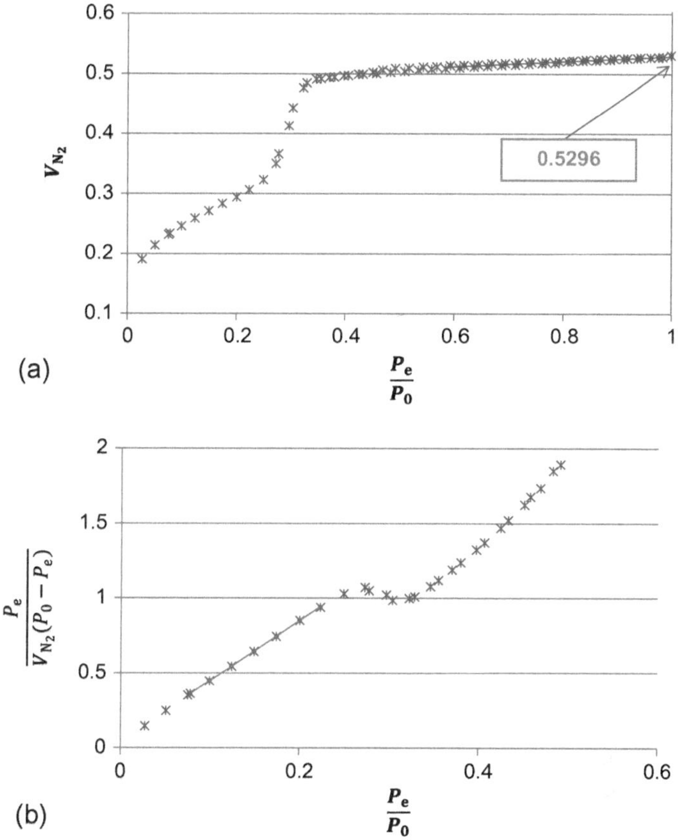

Figure 3.11 Nitrogen adsorption (a) volume adsorption (μl) against P_e/P_0 and (b) BET term $(P_e/V_{N_2}(P_0/P_e))$ as a function of P_e/P_0 for templated silica MCM-41 using N_2 sorption at 77 K.

(b) From Figure 3.11b, we used the lower portion of the isotherm, which is highlighted in blue, to work out the slope and intercept:
- slope = 3.976 (g/μl),
- intercept = 0.0524 (g/μl).

(c) Combining Equations (3.18) and (3.19), the volume of the adsorbed monolayer of N_2 is given:
- $V_m = 0.2482$ μl/g.

(d) From Equation (3.13), the specific surface area A_m is calculated assuming that the thickness of the first packed monolayer of nitrogen[8] is ~0.314-0.40 nm thick:
- $A_m = A_s = \sim 620\text{-}791\ m^2/g$.

(e) Using Equation (3.8) the diameter of the cylinder-shaped MCM-41 is given:
- $d_s = 2r_s = 4 \times (V_s/A_m) = \sim 2.68\text{-}3.41$ nm.

Note: if all nanopores are sphere-shaped, the diameter of the nanopore would be given:

$$\frac{V_s}{A_s} = \frac{n(4/3)\pi r_s^3}{n4\pi r_s^2} = \frac{r_s}{3} \tag{3.20}$$

$$d_s = 2r_s = 6\frac{V_s}{A_s} \tag{3.21}$$

Using Equation (3.21) the diameter of the sphere-shaped MCM-41 is given:
- $d_s = 2r_s = 2 \times (V_s/A_m) = \sim 4.02\text{-}5.12$ nm

3.4 MICROSCOPY

Scanning electron microscopy (SEM) and transmission electron microscopy (TEM) are the visually most informative of analytical techniques because we can 'see' the structural detail of a material under study. They are the most used and well taught subject in all degree courses. Here only the basic principles are restated for the convenience of the readers in order to highlight clearly the differences between SEM, TEM, and the associated elemental X-ray analysis.

3.4.1 SEM, BEM and EDX

Working principles: The useful information obtained from SEM analysis is produced by three different physical processes: images generated from emitted secondary electrons (SE), images from backscattered electrons (BSE), and images from elemental X-ray mapping. SE and BSE are different in their emitting energy and produced by different mechanisms. When a high-energy primary electron interacts with an atom, it undergoes either 'inelastic' or 'elastic' scattering. Inelastic scattering is caused by the interaction of a primary electron with an atomic electron and elastic scattering is the result of an interaction with an atomic nucleus.

During an inelastic collision event, some amount of energy is transferred to another electron. If the energy transfer is very small, the emitted electron may not have enough energy to exit the surface of the material. If the energy transfer is high enough, the emitted electron will exit the solid. An emitted electron with energy less than 50 eV is called an SE. Most SEs seen by SEM are generated within a depth of a few nanometres of the material surface.

In elastic scattering, higher energy electrons are those that have been scattered without loss of kinetic energy by the nucleus of an atom, although collisions may occur after the primary electron has already lost some of its energy to inelastic scattering. A BSE is generally considered to be an emitted electron with energy higher than 50 eV. Most BSEs have energy comparable to the primary beam. *It is worth bearing in mind that the higher the atomic number of a material, the higher the probability that backscattering will occur.* This is why BSE can 'see' multiple phases of a mixture consisting of different materials where SEM cannot.

Elemental characteristic X-ray: In addition to the above two circumstances, the interaction between a primary electron beam and the sample can cause a core electron to be emitted from an atom. The excited atom will relax to its ground state by emitting either a characteristic X-ray or an Auger electron. The emitted X-ray signals can be sorted by their energies using an energy dispersed X-ray (EDX) detector. The primary electrons can travel considerable distances into a solid before losing enough energy through inelastic collisions events. This means that a *smaller* primary beam at any position will produce X-ray emission from a *larger* volume around the position than the beam is focusing on. It is normally not expected to achieve spatial resolution less than ~0.5 μm using this type of analysis.

3.4.1.1 Case Study: drug-loaded sol-gel glass particles

Refer to Figure 2.6—aspirin tablet; Figures 2.9 and 2.10—drug-loaded sol-gel glass particles.

3.4.2 TEM

In TEM, a primary focused beam is incident into a thin film sample. The signal in TEM is obtained from both undeflected and deflected electrons that have penetrated the sample thickness. Penetration of the sample is necessary for TEM analysis and as such the primary beam operates at a much higher voltage than is seen with SEM and EDX, typically in the range of 100-400 keV. The wavelength of 100 keV is only 0.0037 nm, much smaller than that of light and X-rays used in other analytic technologies. So it is expected that TEM has high resolution. For example, the resolution achievable using standard SEM can rarely get below a few nanometres, whereas the TEM can reach sub-nanometre resolution in most circumstances.

TEM can also work as a high-resolution 'X-ray' scattering tool. The diffraction pattern is entirely equivalent to standard XRD (Section 3.3.1).

Taking advantage of this high resolution, single crystals can be seen along certain crystal axes, say orientating with the (0 0 1) axis parallel to the incident electron beam direction, or carrying out pin-point analysis for a polycrystalline thin film.

Limitation: The main drawback with this technique is sample preparation, which is far more complicated than SEM. Samples must first be embedded in a suitable matrix and then sectioned to extremely thin sample slices. Soft or delicate samples can often be damaged during this sampling process.

3.4.2.1 Case Study: drug-loaded mesoporous silica

Refer to Figure 2.11—drug-loaded mesoporous silica.

3.5 TECHNOLOGY FOR EVALUATION OF API AND INORGANIC CRT CARRYING SYSTEM

A significant aspect of the quality of a finished pharmaceutical product is dependent on the quality of the API, which is the manufacturer's overall responsibility. However, during research and product development, it is typically more time and cost effective for the pharmaceutical companies to assess the effectiveness of API quality control. For drug formulations using the inorganic CRT discussed in this book, it is not the intention to list all of the standard methods that pharmaceutical companies follow, but to illustrate a few points that are more relevant to newly developed or improved formulations of inorganic CRT materials for drug delivery.

3.5.1 Evaluation of API and inorganic CRT Carrying System

All of the aforementioned analytical techniques are good tools for assisting the research and development processes during new formulation, and more broadly, new materials development. Some of the methods, such as FTIR and XRD, are routinely used as pharmaceutical quality control test methods and are important techniques for verifying batch release in some cases. In terms of the development of new inorganic CRT formulations though there are two main key questions to resolve:

- Does the formulating process, or residency within the formulation, affect the API integrity?
- Does the new formulation provide an adequate *in vivo* release profile such that the necessary blood serum concentrations are achieved to effect the appropriate therapeutic dosing regimen?

3.5.2 Evaluation of API Purity

API molecules by their very nature tend to be sensitive molecules, prone to some sort of molecular rearrangement under the mildest of conditions, potentially changing the required physicochemical properties of the drug compound in order for it to reach its active site in sufficient concentration. This is in some ways unsurprising. Labile and reactive functionalities are built into API molecules in such a way that they are designed to interact with a biological receptor to induce the intended therapeutic response. An important role of the formulation, or more specifically the excipients, is to protect the API on its journey from dosing to at least the point where it crosses the targeted biological membrane and into the bloodstream. In the development of a new formulation then, it is important to always keep in mind the potential for the formulation materials to effect detrimental physical or chemical change in the API. This is not an easy task, and there are a multitude of ways in which the integrity of a chemical compound can be compromised; hydrolysis, oxidation, thermal degradation, and isomerization are all common potential degradation pathways.[9]

By far the most common method of establishing the presence of impurities or degradation products during product development is by chromatographic separation of all the components coupled with an analyte detection method; see Table 3.3 for a list of the common separation and detection methods.[10]

Each separate method has its own benefits and merits depending on the API in question and the excipients present in the formulation. During the

Table 3.3 Some of the common analyte separation and detection methods used to determine API purity

Separation method	Detection method
LC—liquid chromatography	UV/Vis spectroscopy
HPLC—high-performance liquid chromatography	Mass spectrometry
TLC—thin layer chromatography	NMR—nuclear magnetic resonance
GC—gas chromatography	Fluorescence
CE—capillary electrophoresis	Infrared
CEK—capillary electrokinetic chromatography	
SFC—supercritical fluid chromatography	

development process while any potential impurities and degradation products are unknown, it can be quite a tortuous task finding the right combination of separation and detection methods to best fit the product development. Once formulations reach pilot scale and manufacturing process development, and most of the common impurity and degradation product profiles should be known and identified, the implementation of the most effective yet simplest API assay method is crucial in ensuring efficient QA/QC processing and reliable production batch release. This information and methodology is required for filing pharmaceutical products with the relevant regulatory authorities, so care must be taken to ensure the chosen method is robust and reliable.

3.5.2.1 Case Study: Determination of Genotoxic Impurities in an API Batch

Ritonavir is a protease-inhibiting, antiretroviral drug used in combination with other antiretroviral compounds to treat HIV infection and AIDS. Three compounds (see Figure 3.12a-c) which are known to be potentially genotoxic are used during the chemical synthesis of ritonavir and as such these compounds must be shown to be below the approved safe impurity limits as directed by the FDA and the European Medicines Agency (EMA)—1.5 μg/day.

Many methods have been previously published for the determination of the presence of these impurities individually, at the required limits of detection, but no method had been shown to successfully analyze for the presence of all three impurities concurrently. The authors[11] developed a new method using UPLC (ultra performance liquid chromatography) combined with MS/MS detection, and following this, validated it following the appropriate ICH guidelines. The combination of UPLC-MS/MS resulted in an analysis that was much quicker than any previous method and gave accurate, precise, and robust quantitative determinations of the three genotoxic impurities. The detection limits were low for all three compounds, below 0.3 ppm, as compared with the maximum daily permissible amount of 1.25 ppm.

3.5.3 Toxicity Evaluation

With regard to inorganic CRT formulations, it can be reasonably assumed that the safety and toxicological profile of the API to be formulated is well known. All that is required from that point then is an evaluation and determination of the toxicity profiles of the inorganic CRT materials themselves.

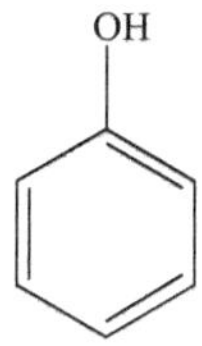

(a)

(b)

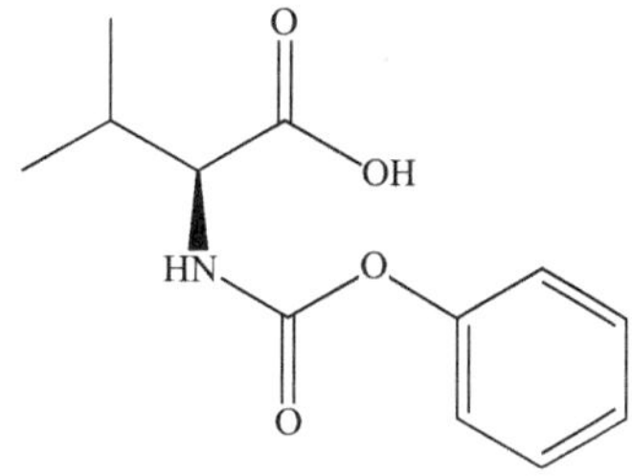

(c)

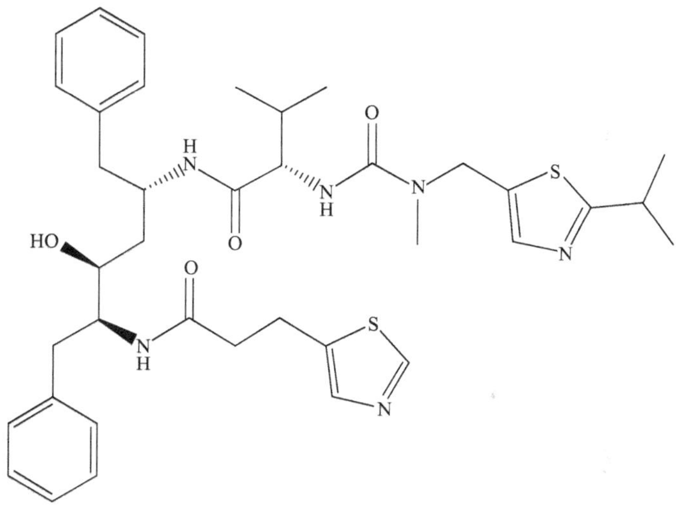

(d)

Figure 3.12 Molecular structures of the three genotoxic impurities: (a) phenol; (b) 4-nitrophenol; (c) *N*-(phenoxycarbonyl)-L-valine; and (d) ritonavir.

The following discussion is based upon the FDA document 'Guidance for Industry—Non-clinical Studies for the Safety Evaluation of Pharmaceutical Excipients'.

According to the FDA,[12] a new excipient is defined as follows:

'any inactive ingredients that are intentionally added to therapeutic and diagnostic products, but that: (1) we believe are not intended to exert therapeutic effects at the intended dosage, although they may act to improve product delivery (e.g., enhance absorption or control release of the drug substance); and (2) are not fully qualified by existing safety data with respect to the currently proposed level of exposure, duration of exposure, or route of administration'.

Based on this definition, it can be seen that the majority of inorganic CRT materials discussed so far (Chapters 1 and 2) and those still to be discussed (Chapters 4-7) can be classified as *excipients*—their purpose is not to impart therapeutic effects, but to modify the release of the drug substance in some way. An exception to this classification may be the family of fusion glasses (discussed in Chapter 1) or the mesoporous bioactive glasses (discussed in Chapter 5), whose design and purpose is to deliver ions through controlled dissolution in a biological setting to induce a therapeutic response. In the eyes of the FDA's guidance it remains to be seen whether a therapeutic ion should fall under the same classification as a drug compound, but in the author's opinion, since the ions are being used to deliver a therapeutic effect directly, then they should be considered as such.

Those inorganic CRT materials that are excipients must be classified further as to whether they are novel or not. The FDA does not refer explicitly to inorganic CRT, but there are many inorganic CRT chemistries already on the GRAS (generally regarded as safe) list or in the Inactive Ingredients Database (IID), meaning they can be generally considered to be an existing excipient since there is clinical data demonstrating safe human use. However, it is important to refer to the administration route and the maximum dosage concentration currently allowed for the product, because if the intended dose is higher or a new administration route is proposed, there may be a need for further toxicological testing. This is largely at the discretion of the regulatory bodies so it is recommended to liaise with them early on to avoid unnecessary testing.

If historical usage data does not exist, then the inorganic CRT material will be classified as a novel excipient and the formulation developers must carry out the industry standard toxicology studies as recommended by the relevant regulatory body and as published in the appropriate ICH guidelines.[13–18]

On the basis of this FDA Guidance document and the materials definition within it, inorganic CRT materials can be split into three categories:

- an active material,
- a new excipient,
- a known excipient.

We will now discuss each of these three categories in turn with the aid of a relevant inorganic CRT material as an example.

3.5.3.1 An Active Material

An active material can be defined as an inorganic CRT material that directly induces a therapeutic effect. An example of this type of material would be the P-glass described in Sections 2.3.6 and 3.2.1 with the intended composition P_2O_5: 40 mol%; CaO: 25 mol%; Na_2O: 5 mol%; MgO: 30 mol%, the MgO content of which was subsequently substituted with SrO. This P-glass was designed to have two interdependent functions, one being a controlled dissolution rate, the other the controlled release of strontium ions. From a materials perspective it can be seen that by changing the amount of strontium in the glass composition in order to modulate the amount of strontium ions released, the dissolution properties of the bulk glass are going to change, which in turn will affect the rate of strontium ion release. It has been demonstrated that the presence of strontium ions in an osseous defect site has an osteogenic effect and stimulates osteoblast cells to accelerate the regeneration of new bone tissue.[19] The purpose of the P-glass in this instance could be to act as a source of controlled release strontium ions in order to improve the rate of bone defect healing.

In this sense the P-glass is acting directly as a therapeutic material, that is, a controlled source of strontium ions and as such would need to be evaluated for its full toxicological profile as stipulated by the FDA and other regulatory bodies.

3.5.3.2 A New Excipient

A new excipient is a material that is not providing any therapeutic function, but has no relevant regulatory authority accepted toxicological information or a history of prior use in humans. An inorganic CRT material that would come under this category would be a mixed metal layered double hydroxide (LDH) (to be discussed in Chapter 7). In the application of these materials, the LDH material is acting purely as a carrier for the drug molecules which are intercalated within the layered structure. The main constituent magnesium hydroxide is an FDA GRAS listed substance so would bypass much of the necessary toxicology testing, but the mixed hydroxides where extra metal ions are introduced to the magnesium hydroxide tetrahedral network during the material synthesis, do not feature. These materials would require full toxicity testing as required by the appropriate pharmaceutical products regulatory authority.

3.5.3.3 A Known Excipient

These materials can be defined as those that are inactive in their role in pharmaceutical formulations and have documented safe use for human consumption. A material listed on the FDA's GRAS database would qualify for this status. Non-ordered porous silica would feature as an inorganic

CRT material in this category since its role in inorganic CRT formulations is only to modulate the rate of drug release from within its porous structure and is not known to induce any sort of therapeutic, or in fact biological, response at all when taken orally. The broader family of silicon dioxides are regularly used as additives in the food industry and the naturally sourced variety diatomaceous earth, arising from the fossilized remains of diatoms, has found extensive use as a filter aid in the food industry due to its natural, well-ordered porous structure. The LD_{50} of another related silicon dioxide, Perlite, has been shown to be in excess of 10 g/kg in a rat model. Perhaps more prevalent in this context is the widespread use of silicon dioxide, in particular colloidal silica, in drug formulations as an anticaking agent, a tablet disintegrant, and a viscosity-increasing agent.

The intention of this final section of this chapter was not to detail the necessary toxicology testing required to gain approval for new materials, but to provide a guide to the reader in order that they can make an informed, early decision on which regulatory pathway they need to consider when in the process of developing new inorganic CRT materials. The researcher is encouraged to engage with the relevant body early on in the formulation development process such that the required testing can be built into the project pipeline in a timely fashion.

REFERENCES

1. Clarke MT, Lee PTH, Arora A, Villar RN. Levels of metal ions after small- and large diameter metal-on-metal hip arthroplasty. *J Bone Joint Surg [Br]* 2003;**85-B**:913–7.
2. Janin A, Maache M, Lavalley JC. FTIR study of the silanol groups in dealuminated HY zeolites: nature of the extraframework debris. *Zeolites* 1991;**11**(4):391–6.
3. Serra J, Gonzalez P, Liste S, Serra C, Chiussi S, Leon B, et al. FTIR and XPS studies of bioactive silica based glasses. *J Non-Cryst Solids* 2003;**332**:20–7.
4. Raskar D, Rinke MT, Eckert H. The mixed-network former effect in phosphate glasses: NMR and XPS studies of the connectivity distribution in the glass system $(NaPO_3)_{1-x}(B_2O_3)_x$. *J Phys Chem C* 2008;**112**(32):12530–9.
5. Markovic M, Fowler BO, Tung MS. Preparation and comprehensive characterization of a calcium hydroxyapatite reference material. *J Res Natl Inst Stan* 2004;**109**(6):553–68.
6. Brunauer S, Emmett PH, Teller E. Adsorption of gases in multimolecular layers. *J Am Chem Soc* 1938;**60**(2):309–19.
7. De Lange MF, Vlugt TJH, Gascon J, Kapteijn F. Adsorptive characterization of porous solids: error analysis guides the way. *Microporous Mesoporous Mater* 2014;**200**:199–215.
8. http://www.getnitrogen.org/pdf/graham.pdf [accessed 09.04.15].
9. Fathima N, Mamatha T, Qureshi HK, Anitha N, Rao JV. Drug-excipient interaction and its importance in dosage form development. *J Appl Pharm Sci* 2011;**1**(6):66–71.
10. Singh S, Handa T, Narayanam M, Sahu A, Junwal M, Shah RP. A critical review on the use of modern sophisticated hyphenated tools in the characterization of impurities and degradation products. *J Pharmaceut Biomed* 2012;**69**:148–73.

11. Venugopal N, Reddy AVB, Madhavi G. Development and validation of a systematic UPLC-MS/MS method for simultaneous determination of three phenol impurities in ritonavir. *J Pharmaceut Biomed* 2014;**90**:127–33.
12. FDA, Guidance for industry—nonclinical studies for the safety evaluation of pharmaceutical excipients 2005. Additional copies from: office of training and communications division of drug information, HFD-240, Center for Drug Evaluation and Research, Food and Drug Administration, 5600 Fishers Lane Rockville, MD, http://www.fda.gov/cder/guidance/index.htm.
13. ICH guidance for Industry S7A safety pharmacology studies for human pharmaceuticals.
14. ICH guidance for industry M3 nonclinical safety studies for the conduct of human clinical trials for pharmaceuticals, http://www.fda.gov/cder/guidance/index.htm.
15. ICH guidelines for Industry S3A Toxicokinetics: the assessment of systemic exposure in toxicity studies and s3b pharmacokinetics: guidance for repeated dose tissue distribution studies, http://www.fda.gov/cder/guidance/index.htm.
16. ICH guidance for industry S2B genotoxicity: a standard battery for genotoxicity testing of pharmaceuticals, http://www.fda.gov/cder/guidance/index.htm.
17. ICH guidelines for industry S5A detection of toxicity to reproduction for medicinal products and s5b detection of toxicity to reproduction for medicinal products: addendum on toxicity to male fertility, http://www.fda.gov/cder/guidance/index.htm.
18. ICH guideline for industry S1A the need for long-term rodent carcinogenicity studies of pharmaceuticals, http://www.fda.gov/cder/guidance/index.htm.
19. Wu C, Chang J. Multifunctional mesoporous bioactive glasses for effective delivery of therapeutic ions and drug/growth factors. *J Control Release* 2014;**193**:282–95.

FURTHER READING

Brundle C, Evans Jr CA, Wilson S. *Encyclopedia of materials characterization*. Shelter Island: Manning Publications Co; 1992.

CHAPTER 4

Silica-Based Amorphous Drug Delivery Systems

Contents

4.1 Introduction 93
4.2 Sol-Gel Chemistry and Processing 94
4.3 Ordered Mesoporous Silica 98
4.3.1 Block copolymer templated systems 99
4.3.1.1 Hildebrand solubility parameter 99
4.3.1.2 Meaning of the Hildebrand solubility parameter 100
4.3.1.3 Measurement and estimation of solubility parameters 101
4.3.1.4 Flory-Huggins theory 101
4.3.1.5 Meaning of the Flory-Huggins interaction parameter 103
4.3.1.6 Relationship between the Flory-Huggins interaction parameter and solubility parameters 103
4.3.1.7 Ordered polymer template systems 104
4.3.1.8 Phase separation of block copolymer 104
4.3.1.9 The physical meaning of $N\chi_{AB}$ *105*
4.3.1.10 Designing ordered polymer templates using block copolymers 106
4.3.2 Ordered mesoporous silica *via* emulsion chemistry 112
4.3.2.1 Non-ionic surfactant templates 114
4.3.2.2 Ionic surfactant templates 117
4.4 Non-Ordered Porous Silica for Controlled Release 120
4.4.1 Porous silica by precipitation chemistry 122
4.4.2 Non-ordered silica by the sol-gel chemistry of silicon alkoxides 127
4.4.3 Processing of non-ordered porous silica 128
4.4.3.1 The contribution of matrix dissolution to drug release 130
4.5 Porous Silica for Improved Efficacy and Tolerability of Topically Administered Active Compounds 131
References 134

4.1 INTRODUCTION

Being the second most abundant element on the Earth, and as it features in the chemistry of such a wide range of technologies and fields of science and engineering (traditional glass and ceramic materials through to cutting edge

Inorganic Controlled Release Technology
http://dx.doi.org/10.1016/B978-0-08-099991-3.00004-1

semi-conductor applications), it is perhaps no surprise that such a prominent field of nanotechnology research for controlled release applications centres around the simple silicon tetrahedron.

The breakthrough for the use of silicate materials for healthcare applications came in the late 1960s when it was found that a specific composition of silicate glass, formed *via* a traditional melt-quench process, was found to form a surface hydroxyapatite layer *in vivo* which then formed strong chemical bonds to collagen.[1] This silicate-based glass, later more commonly known as Bioglass®, was the first such material shown to have bioactive properties.

However, formation of glassy materials by fusion of metal oxides and quenching from above their liquidus temperatures results in amorphous materials that are dense and non-porous. Research into the area of bioactive glasses subsequently increased hugely and a number of novel inorganic materials were found to have potential application as biomedical materials. In particular, research focussed on a method of forming porous metal oxide materials known as sol-gel chemistry. Sol-gel chemistry had been well known for a number of years, the first reported porous silica gels made *via* sol-gel processing in 1846 by M. Ebelmen at the 'Manufacture de Ceramiques de Sevres' in France.[2,3] However, it was only in 1981 when the concepts of preparing bioactive glasses in combination with sol-gel processing to form porous silicate materials were internationally first recognized and the predecessor materials to general inorganic controlled release technologies were first realized.

4.2 SOL-GEL CHEMISTRY AND PROCESSING

The sol-gel reactions of silicate precursors are of such importance to the production of porous silicate materials for inorganic CRT applications that some of the fundamental theories must be introduced here such that the reader is well placed to understand how these materials can not only be made but also be designed to be fit for purpose (for a much more in-depth treatise on the sol-gel reactions of silicon alkoxides and other metal alkoxides, please see Brinker and Scherer, Sol-Gel Science, 1990, Academic Press Inc.).

There are many benefits to the production of amorphous materials *via* sol-gel processing rather than by melt-fusion processes. There are obvious energy and processing benefits to be had as sol-gel processing is carried out at or near to room temperature, as opposed to the high temperatures required to melt silicate glasses (typically >1200 °C). This low-temperature processing also provides access to certain compositions of glass which are difficult to achieve because of very high melting temperatures. The products of sol-gel processing tend to be of high purity and homogeneity, and in the context of controlled release applications, most importantly they are porous.

At its most basic level, formation of silicate networks by the sol-gel process involves the hydrolysis and condensation reactions of silicon alkoxide materials such as the tetra-substituted silicon alkoxides tetramethylorthosilicate (TMOS, $Si(OCH_3)_4$) and tetraethylorthosilicate (TEOS, $Si(OCH_2CH_3)_4$), which provide the necessary silicon centres. Hydrolysis of the alkoxide ligands attached to the silicon metal centre opens up the possibility of condensation reactions between silicon monomers, eventually forming an amorphous polysilicate network.

$$Si(OR)_4 \rightarrow SiO_{2(\text{amorphous})} \tag{4.1}$$

This polymerization process occurs in three discrete stages[4]:

1. polymerization of monomer units to form colloidal particles,
2. growth of particles,
3. the particles then link to form chains and then networks which extend throughout the reaction medium, eventually thickening it to a gel.

To break this process down into the processes occurring at the molecular level, three reversible chemical reactions take place driven by the equilibrium of the chosen reaction system:

$$(OR)_3Si\text{-}OR + H_2O \underset{\text{esterification}}{\overset{\text{hydrolysis}}{\rightleftharpoons}} (OR)_3Si\text{-}OH + ROH \tag{4.2}$$

$$(OR)_3Si\text{-}OR + HO\text{-}Si(OR)_3 \underset{\text{alcoholysis}}{\overset{\text{alcohol condensation}}{\rightleftharpoons}} (OR)_3Si\text{-}O\text{-}Si(OR)_3 + ROH \tag{4.3}$$

$$(OR)_3Si\text{-}OH + HO\text{-}Si(OR)_3 \underset{\text{hydrolysis}}{\overset{\text{water condensation}}{\rightleftharpoons}} (OR)_3Si\text{-}O\text{-}Si(OR)_3 + H_2O \tag{4.4}$$

Silicon alkoxides are immiscible with aqueous solutions and as a result organic cosolvents are often used to homogenize the initial reaction systems. The simultaneous use of organic solvents is not essential however as hydrolysis of the silicon alkoxide across the phase boundaries generates alcoholic solutions *in situ* which are effective in homogenizing the reaction mixture as it proceeds.

The hydrolysis of the silicon alkoxide precursors does take place under neutral conditions, but the rates of these reactions are slow enough to make the use of a catalyst essential. Mineral acids such as hydrochloric, sulphuric and nitric acid are most commonly used, as well as ammonium hydroxide, but in principle any compounds demonstrating suitable acidic or basic character may be used.

The nature of the products formed during the initial stages of hydrolysis and condensation is very much dependent on the pH of the reaction system and is generally referenced to the isoelectric point of silica at pH ~2 (Figures 4.1 and 4.2). The reaction pH is a key factor in determining the time taken for the reaction system to gel, as can be seen in Figure 4.3.

As such it can be seen that above the isoelectric point of silica, the condensation rate is proportional to $[OH^-]$. Fast deprotonation of the most acidic silanol (Si—OH) groups takes place most favourably, followed by a slow condensation reaction with another labile silanol functional group.

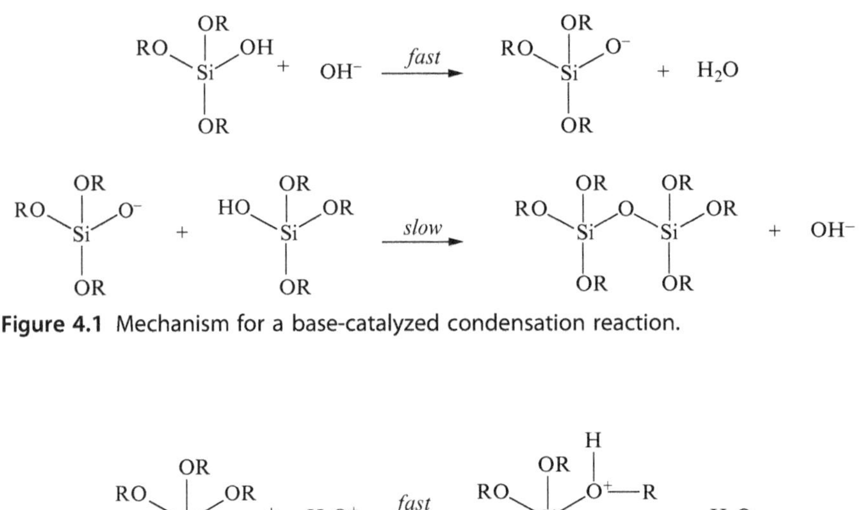

Figure 4.1 Mechanism for a base-catalyzed condensation reaction.

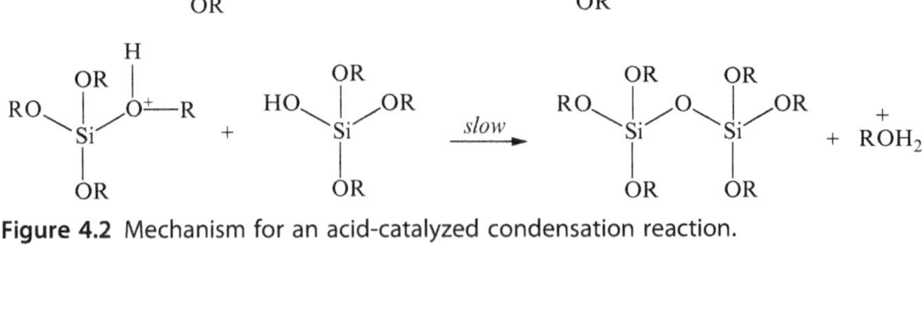

Figure 4.2 Mechanism for an acid-catalyzed condensation reaction.

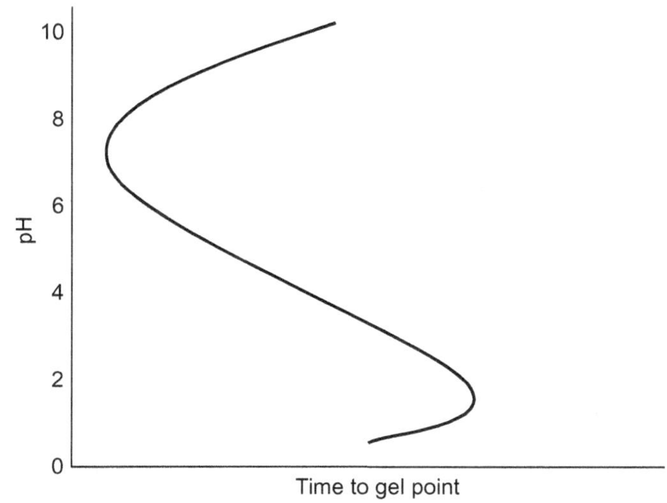

Figure 4.3 Gel time versus pH for aqueous silicates.

The most acidic silanols are those on the most highly condensed species; therefore, condensation reactions take place most favourably between larger charged species and smaller neutral species. Particle growth therefore occurs by the addition of low-molecular weight species to highly condensed species after which aggregation of the large condensed species takes place forming chains and networks. In this pH range, the solubility of silica is low and as the primary particles reach ~2-4 nm in size their growth stops.

Similarly, below pH 2 the rate of gel formation is proportional to $[H^+]$ (H_3O^+).

In this pH range, as the silicate species present are not highly ionized, particle growth and aggregation occur simultaneously and the gel networks formed are composed of very small primary particles.

Above pH 7, polymerization occurs in the same manner as discussed previously between pH 2 and 7. However at these higher pH levels, a greater number of the silicate particles are likely to be negatively charged and as such mutually repulsive. This electrostatic repulsion prevents particle aggregation, whilst particle growth occurs by the addition of smaller neutral species. Depending on the presence of other ionic species present, stable sols containing particles of up to 100 nm can be formed very quickly.

The molar ratio of water:silicon alkoxide also has an important role to play in determining the properties of the silicate glasses formed by the sol-gel process. Since water is produced as a by-product of the reaction processes involved, it is theoretically possible to fulfil the conditions of Equation (4.1) by using only two molar equivalents of water. In practice, however, even when large excesses of water are used the hydrolysis of the silicon alkoxide starting materials does not go to completion before the onset of condensation reactions and a range of intermediate species are typically formed. As an example of the effects of varying this process parameter, when 20–50 equiv. of water are used under basic conditions, monodisperse, spherical particles are produced.[5] In contrast, a two-step hydrolysis procedure under acidic conditions (1 equiv. of water followed by a further 3-4 equiv. of water) produced ramified, weakly branched products.[6]

Many of the other process variables in the sol-gel reaction can be varied to produce amorphous silicate products with varying physical properties. Discussion of their control and the direct effects of each in turn are beyond the scope of this text. The reaction pH and the water:silicon alkoxide ratio are two of the most significant, but at the fundamental level the relative kinetics and the balance between the rates of the hydrolysis and condensation reactions of the silicon monomers are crucial in determining the properties of sol-gel-derived silica products.

4.3 ORDERED MESOPOROUS SILICA

Due to the rapidly increasing importance of CRT for active delivery, a significant amount of research effort has focussed on developing drug delivery systems that are able to deliver increasingly larger dosages. Porous silicates emerged as early preferred choices for inorganic CRT because of their excellent chemical, physical, thermal and mechanical stability as well as chemical versatility and biocompatibility.

This ground breaking technology was first identified in the early 1990s when the mesoporous silicates MCM-41[7,8] and SBA silica[9] were first developed. The unique properties of mesoporous silica include high surface area (>900 $m^2\ g^{-1}$), large pore volume (>0.9 $cm^3\ g^{-1}$), and tuneable pore size with a narrow distribution (2-10 nm). The idea to use surfactants acting as templates to develop well-ordered and regulated nanoporous structures immediately attracted worldwide attention. It is still a new era of technology and research and is expected to develop into a new branch of nanotechnology with the purpose of preparing mesoporous materials.

However, the science, or nanoscience, to study highly ordered mesoporous materials is not yet well developed. Due to their high surface areas and pore volumes, ordered mesoporous structures are able to absorb/encapsulate relatively large amounts of bioactive molecules and as such will have great potential for inorganic CRT applications. What follows are some of the basic theories and strategies necessary in order to design and formulate ordered porous silicate matrices for controlled release applications.

4.3.1 Block Copolymer Templated Systems

The use of block copolymers as structure-directing agents is one of the most important recent developments for the preparation of nanostructured hybrid materials. From a materials perspective, block copolymers are ideally suited to the development of advanced hybrid materials. Their use in the formation of porous silicate materials can provide compositional and morphological control down to the nanometre scale, which will certainly aid in the design and development of a range of mesoporous silicas. In this section, polymer solution theories and thermodynamic principles will be first introduced. Some concepts and materials parameters will also be discussed for improved understanding of the theories and applications and to provide insights into the mechanisms of forming mesoporous structures and to achieve compositional and morphological controls.

4.3.1.1 Hildebrand Solubility Parameter

The concept of solubility parameters[10] was first introduced by Charles M. Hansen in 1967 in his PhD thesis to evaluate whether one material will dissolve in another and form a solution. The concept is applicable to all materials when miscibility and/or compatibility between different materials are concerned and is based on the idea that 'like dissolves like'. It is normally true that one molecule would 'like' another if they have similarities in their chemical structure and properties. Of significance is the fact that the solubility parameter provides a numerical estimate of the degree of interaction between materials. Solubility parameters have often been used as a good indication of *solubility*, and are regularly used to predict whether two materials will mix efficiently (like each other) or not (dislike each other).

The Hildebrand solubility parameter δ is defined as the square root of the cohesive energy density. The cohesive energy density is the amount of energy needed to completely remove a unit volume of molecules from their neighbours to infinite separation. This is equal to the heat of vaporization of the compound divided by its volume in the condensed phase and is given by

$$\delta = \frac{E}{V} = \sqrt{\frac{\Delta H_{\mathrm{v}} - RT}{V}} \tag{4.5}$$

where E is the cohesive energy, V the volume, ΔH_v the vaporization enthalpy, R the gas constant and T is the temperature.

Hildebrand and Scott[11] also provided a model to establish the relationship between the solubility parameter and the mixing enthalpy ΔH_m

$$\Delta H_m = (\delta_A - \delta_B)^2 \phi_A \phi_B V_m \tag{4.6}$$

where V_m is the mixture volume, ϕ is the volume fraction and subscripts A and B represent materials A and B, respectively.

The concept was originally thought of as an application for non-polar materials such as polymers. The solubility parameters are very useful material properties for the design and development of nanostructure templates when block polymers, and other materials, are used. Solubility parameters are applicable to any material or mixture of materials regardless of their physical states – liquid, solid, or something in between.

4.3.1.2 Meaning of the Hildebrand Solubility Parameter

The physical meaning of the solubility parameter is a measure of the interactions between different materials at the molecular level. In general, we can classify three types of interaction forces. The first, most common, types of interactions are called the London dispersion forces, which are particularly relevant to dispersive interactions between materials. All molecules have this type of attractive force arising from moving negatively charged electrons orbiting around central positively charged nuclei. The moving negative charges create transient electromagnetic fields which attract all atoms to each other with no directional preference. Equations (4.5) and (4.6) were originally derived from this mechanism which was thought to be only applicable to non-polar materials.

The second types of interactions are polar cohesive forces produced through permanent dipole-dipole interactions. They are inherently molecular interactions in molecules where dipole-dipole interactions exist.

The third type of interaction is hydrogen bonding, which is well known. The key differences between hydrogen bonding and the other two forces are its directional preference, one-to-one specificity and the magnitude of the resulting interaction force, a hydrogen bond can be in the range 5-30 kJ mol^{-1} where dipole-dipole and non-polar interactions are typically less than 1 kJ mol^{-1}. So to account for all possible interactions, the cohesive energy can be generally expressed to incorporate all three components:

$$E = E_D + E_P + E_H \tag{4.7}$$

And the corresponding solubility parameter of the three types of interactions:

$$\delta^2 = \delta_D^2 + \delta_P^2 + \delta_H^2 \tag{4.8}$$

where subscripts D, P and H stand for dispersion, polar and hydrogen bonding.

4.3.1.3 Measurement and Estimation of Solubility Parameters

For solvents, their solubility parameters can be calculated through measuring heat capacity ΔH_v (Eq. 4.5). There is no need to measure these parameters because almost all solubility parameters of solvents have been previously established and are available in many published books. The reference 'Polymer Handbook'[12] collects many of the commonest solvents and their corresponding solubility parameters.

It is however impossible to measure the heat capacity of a polymer in order to calculate its solubility parameter because it cannot be evaporated. To solve this problem, all polymer solubility parameters are measured indirectly using the solubility parameters of solvents. The basic idea is to employ a series of solvents, establish which one is the most effective at solubilizing the particular polymer and then assign the solubility parameter of the solvent to the polymer. Commonly used methods include solvent screening, polymer swelling, refractive index and intrinsic viscosity. It is expected that polymer parameters obtained through indirect methods are only estimated values, and it is therefore not surprising that there are differences found in literature-reported solubility parameters.

4.3.1.4 Flory-Huggins Theory

The Flory-Huggins theory is used to assess the miscibility of polymer blends and is based upon statistical thermodynamic models. It was developed independently by Flory[13,14] and Huggins[15,16] in the 1940s and as such the corresponding thermodynamic theories of polymer solutions and/or blends were jointly termed as the Flory-Huggins theory. Different to the above Hildebrand Solubility Parameter, it provides fundamental understanding backed with classical thermodynamic theories.

The basic thermodynamic relationship governing mixtures is given by Gibbs free energy ΔG_m. The relationship between ΔG_m and mixing enthalpy and entropy at temperature T is given by

$$\Delta G_m = \Delta H_m - T\Delta S_m \tag{4.9}$$

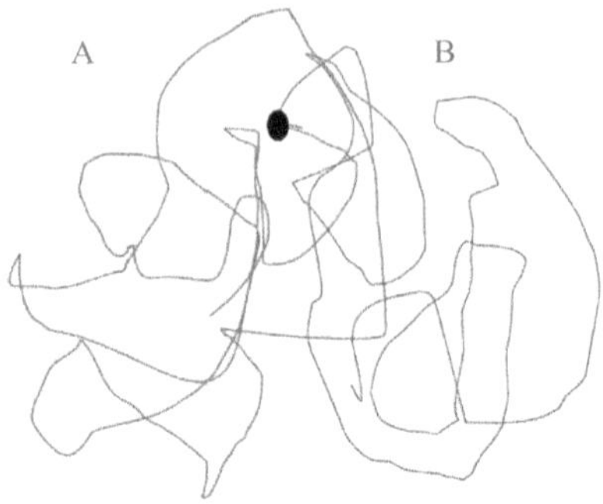

Figure 4.4 Schematic plot of a binary block copolymer.

Let us use a binary system as an example to discuss the Flory-Huggins theory, from which several important parameters will be introduced. Figure 4.4 is a schematic plot showing a block copolymer chain consisting of two blocks, A and B. The two blocks are joined together by a covalent bond as highlighted by the black dot shown in Figure 4.4. Both of the polymer chains are assumed to stay in their natural coiled status/conformation.

Flory-Huggins worked out the mixing entropy change ΔS_m:

$$\Delta S_m = -\kappa\left(\frac{\phi_A}{N}\ln\phi_A + \frac{\phi_B}{N}\ln\phi_B\right) \tag{4.10}$$

and the mixing enthalpy change ΔH_m:

$$\Delta H_m = \kappa T \phi_A \phi_B \chi_{AB} \tag{4.11}$$

where Flory-Huggins defined an interaction parameter, χ_{AB}:

$$\chi_{AB} = \left(\frac{z}{N_B T}\right)\left\{\varepsilon_{AB} - \frac{1}{2}(\varepsilon_{AA} + \varepsilon_{BB})\right\} \tag{4.12}$$

where κ is the Boltzmann constant, $N = (N_A + N_B)$ the number of repeating monomer units in the copolymer, ϕ_A and ϕ_B are the volume fractions of polymers A and B, z is the coordination number (the number of nearest neighbours surrounding a given interaction unit), ε_{AB}, ε_{AA} and ε_{BB} are the interaction energies per repeating unit between A and B, A and A and B and B, respectively.

The complete Flory-Huggins theory is obtained by combination of Equations (4.9)–(4.11) giving:

$$\frac{\Delta G_m}{\kappa T} = \frac{\phi_A}{N}\ln\phi_A + \frac{\phi_B}{N}\ln\phi_B + \phi_A\phi_B\chi_{AB} \tag{4.13}$$

Equation (4.13) is the theory in terms of the free energy change upon mixing.

As long polymer chains have fewer mixed configuration and conformation states, the entropic contribution in the first two terms in Equation (4.13) will be approaching zero with increasing molecular weight of the block A and B for the binary system. This is because the degree of polymerization, or repeating unit N, is normally a very big number. Therefore, the phase behaviour of the block copolymer is then largely determined by the value of χ_{AB}, that is the Flory-Huggins interaction parameter.

4.3.1.5 Meaning of the Flory-Huggins Interaction Parameter

As defined in Equation (4.12), the meaning of the χ_{AB} is the energy change after mixing two polymer chains A and B. This energy change should be considered a result of the polymer chain interactions between molecules or smaller subunits of the two different polymer chains. For convenience, it normally uses the polymer repeating unit as a basic structural unit to count for the interactions between the same and different polymer chains. The factor that decides the interaction is chemical structure, which in turn determines the χ_{AB} value. The interaction energies between blocks A and B, A and A and B and B are ε_{AB}, ε_{AA} and ε_{BB}, respectively, and all contribute to the Flory-Huggins interaction parameter χ_{AB}. In addition, the number of nearest neighbours surrounding a given interaction unit also affects χ_{AB}, which is the coordination number z in Equation (4.12).

It is worth noting that the most significant feature of the Flory-Huggins interaction parameter χ_{AB} is that it is a measureable, physical property based on thermodynamic theory.

4.3.1.6 Relationship between the Flory-Huggins Interaction Parameter and Solubility Parameters

Combination of Equations (4.6) and (4.11), to the first approximation, means that the Flory-Huggins interaction parameter can be expressed by

$$\chi_{AB} = \frac{(\delta_A - \delta_B)^2 V_m}{\kappa T} \tag{4.14}$$

The unit mixture volume can be expressed by

$$V_m = \frac{M}{N_a \rho} \tag{4.15}$$

where ρ and M are the material density and molecular weight of repeating unit; N_a is the Avogadro constant.

Substituting Equation (4.15) into Equation (4.13) gives

$$\chi_{AB} = \frac{(\delta_A - \delta_B)^2}{\kappa T}\frac{M}{\rho N_a} = \frac{M}{\rho RT}(\delta_A - \delta_B)^2 \tag{4.16}$$

Therefore, Flory-Huggins interaction parameters χ_{AB} can be estimated by using solubility parameters to the first approximation.

4.3.1.7 Ordered Polymer Template Systems

The self-assembling properties of block copolymers provide a range of unique platforms for the design and development of nanostructured materials. Block copolymers consist of two or more chemically distinct polymer blocks covalently bonded together. These blocks can be thermodynamically incompatible with each other as the entropy of mixing per unit volume is small and varies inversely with molecular weight. In the bulk, binary copolymers separate at the molecular level to produce nanostructures with compositional and morphological variations. By changing chemical structure, molecular weight and architectures with the aid of different solvent systems, a range of new ordered polymer templates can be developed. This is a new branch of nanotechnology with great potential in the future for controlled release technology as well as for many other applications.

4.3.1.8 Phase Separation of Block Copolymer

New development of nanostructured hybrid materials has shown that block copolymers play important roles. Phase separation at the nanometre scale provides a unique opportunity to form nano-ordered mesoporous structural materials. A wide range of choices for different polymer systems will enable scientists to create a multitude of well-ordered nanotemplates for a wide range of applications.

Let us use the binary block copolymer as an example to study thermodynamics using the Flory-Huggins theory. Equation (4.13) is used to predict the possibility and trend of phase separation for a given ideal block copolymer. By rearranging Equation (4.13) it can be transformed to another form:

$$\frac{N\Delta G_m}{\kappa T} = \phi_A \ln\phi_A + \phi_B \ln\phi_{AB} + N\phi_A\phi_B\chi_{AB} \tag{4.17}$$

Figure 4.5 is a plot made according to Equation (4.17) by plotting ϕ_A against $(N\Delta G_m/\kappa T)$ as a function of $N\chi_{AB}$, where seven values of $N\chi_{AB}$ have been highlighted in the figure to aid analysis, increasing from 0, 1, 2 and up to 6.

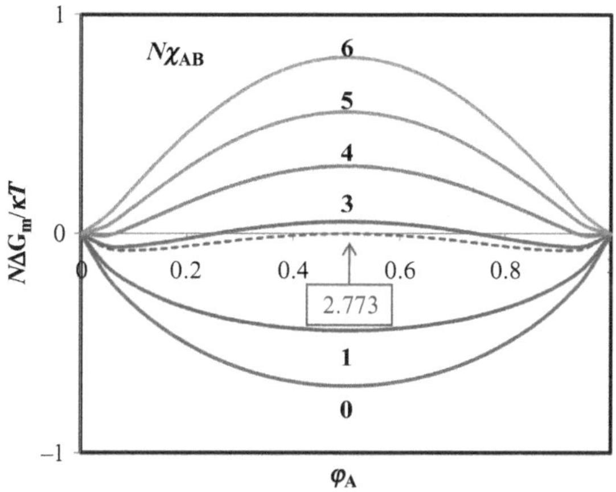

Figure 4.5 Gibbs free energy as a function of composition ϕ_A and interaction parameter $N\chi_{AB}$.

4.3.1.9 The Physical Meaning of $N\chi_{AB}$

It is worth considering the importance of $N\chi_{AB}$. χ_{AB} is defined by Flory-Huggins as an interaction parameter between two different polymers, A and B, at a molecular level. N is the degree of polymerization which is taken as the total number of repeating units of the polymer. Therefore, $N\chi_{AB}$ is the interaction per polymer.

When $N\chi_{AB}=0$, which can be treated as a special case, $\delta_A=\delta_B$ (refer to Equation (4.16)). Blocks A and B are virtually the same and should be 100% miscible and compatible.

In general, when $N\chi_{AB}\leq 2.773$ (the dotted line in Figure 4.5) $(N\Delta G_m/\kappa T)\leq 0$, the system will remain as a single phase with no phase separation occurring at all concentrations ϕ_A 0 up to 1.

When $N\chi_{AB}\geq 2.773$: $N\Delta G_m/\kappa T$ can be less than zero or greater than zero depending on the concentration of blocks A and B in the copolymer. The general trend is that with increasing $N\chi_{AB}$ above 2.773, the probability of keeping $N\Delta G_m/\kappa T<0$ becomes smaller and smaller. At $N\chi_{AB}\geq 4$, $N\Delta G_m/\kappa T$ is greater than zero for almost all concentrations in the region ϕ_A 0 up to 1. This indicates that the binary polymer system will coexist as a two-phase system, that is block polymer A will form one phase and block polymer B another.

There are two particular ranges where it can be predicted with confidence whether phase separation will occur or not. When $N\chi_{AB}<2.773$

no phase separation will occur and at $N\chi_{AB} > 4$ the two constituent polymers will phase separate. From these two important boundary values the critical degree of polymerization, or average repeating unit N_c can be worked out. When $N_c < 2.773/\chi_{AB}$ a single-phase system occurs and when $N_c > 4/\chi_{AB}$ a two-phase system occurs.

Figure 4.5 shows how the variation of Gibbs free energy ΔG_m depends on two parameters, one is $N\chi_{AB}$ and the other is ϕ_A. This means that the interaction energy $N\chi_{AB}$ and the concentration of block A or block B (i.e. the ratio of block A chain length to block B) together can decide if the polymer binary system will exist as a one or two-phase system preferentially.

Under most circumstances, polymer blocks A and B are most likely to adopt states of phase separation. This is because the degree of polymerization is normally in the range of a few hundreds to over 1000. Even for a small value of χ_{AB}, the product of $N\chi_{AB}$ would normally become a big number, which can easily be greater than 2.773 in most circumstances. This is why most polymer blends tend to phase separate. This can be seen from Figure 4.5 where the bigger the value of $N\chi_{AB}$, the greater the driving force towards phase separation in terms of $N\Delta G_m/\kappa T$.

4.3.1.10 Designing Ordered Polymer Templates Using Block Copolymers

Figure 4.5 shows there is a high tendency of phase separation for a binary block copolymer system. The thermodynamic theory predicts that the two polymeric blocks A and B are in favour of phase separation to reduce the free energy of the system. However, the two parts of the block copolymer are covalently bonded to each other (refer to Figure 4.4). Macro-phase separation cannot occur because it would be stopped by the bond that holds the two different polymer chains together permanently. To minimize the interfacial energy between the two polymer chains, the force that drives the phase separation will make the polymer chain stretch away from its natural conformation. For example, the chain segments of blocks A and B will transform from coiled chain states to more stretched states in order for block A to move away from block B and *vice versa*. The scale and dimension of the polymer chains stretching away from each other depend on many factors, including the polymer chain length of block polymers A and B, the interaction parameter χ_{AB}, solubility parameter δ and ultimately, their chemical structures and associated configuration and conformation. All of these variables have provided scientists with many choices in the design and development of different nanostructures and architectures of block copolymers. It is worth bearing in mind that all the interactions between block polymers A and B are considered at the molecular level so phase separation also operates at molecular and/or nanometre scales.

It is well known that binary block copolymers can form well-ordered nanostructures. A good example is the copolymer of polystyrene-b-polyisoprene, that is, PS-b-PI. The solubility parameters of PS and PI are 9.2 and 8.1 (cal $cm^{-3})^{1/2}$, respectively.[12] By using Equation (4.12), to the first approximation, the Flory-Huggins interaction parameter is $\chi_{AB}=0.086$, giving critical degree of polymerization values (N_c) of 32 and 47 for the two critical conditions of phase homogeneity and phase separation, respectively (assuming that the average density is 1.01 g cm^{-3} and the average molecular weight of the repeating unit of the two blocks is 172 g mol^{-1}). For a normal PS-b-PI copolymer, the degree of polymerization is much greater than 47 so phase separation between the two polymer blocks always occurs. The morphology changes with PS and PI compositional variation as follows:

$\phi_{PS}<0.15$	Sphere (S)—PS dominating
ϕ_{PS} 0.15-0.35	Cylinder (C)
ϕ_{PS} 0.35-0.65	Lamellar (L)
ϕ_{PS} 0.65-0.85	Cylinder (C′)
$\phi_{PS}>0.85$	Sphere (S′)—PI dominating

Based on the above discussion and the analysis of $N\chi_{AB}$ and ϕ_A and phase separation around Figure 4.5, a more generalized morphological map of a binary block copolymer can be drawn using the two factors, $N\chi_{AB}$ and ϕ_A. Figure 4.6 is the morphological map of drawing $N\chi_{AB}$ as a function of ϕ_A. There are two dotted lines. The bottom one is the lower limit of $N\chi_{AB}$, below which no phase separation occurs but a single-phase system exists, that is, blocks A and B are miscible. The other dotted red line is an upper limit, above which no single phase can form, only binary phases, from concentration ϕ_A equal to 0 up to 1. For a given block copolymer, composition variation of ϕ_A will decide the morphology. The dimension of each morphological phase must be very small. These phases should normally be at the nanometre scale because they are much smaller than fully stretched polymer chains, but bigger than the dimension of their natural coiled states. Figure 4.7 is a schematic drawing to show one example of phase separation of blocks A and B in a monolayer

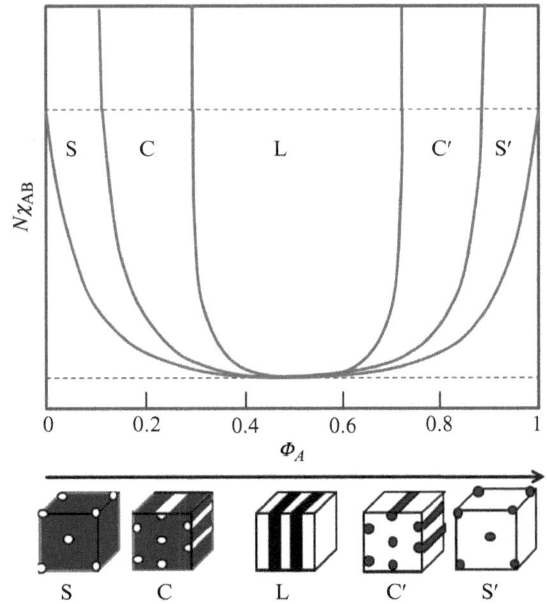

Figure 4.6 Theoretic prediction of phase diagram of binary block polymer as a function of the interaction parameter $N\chi_{AB}$ and composition in volume fraction ϕ_A. (*Note*: there are two more cubic discontinuous gyroid G and G′. They should sit between C and L and L and C′ respectively, which are not shown here for simplicity.)

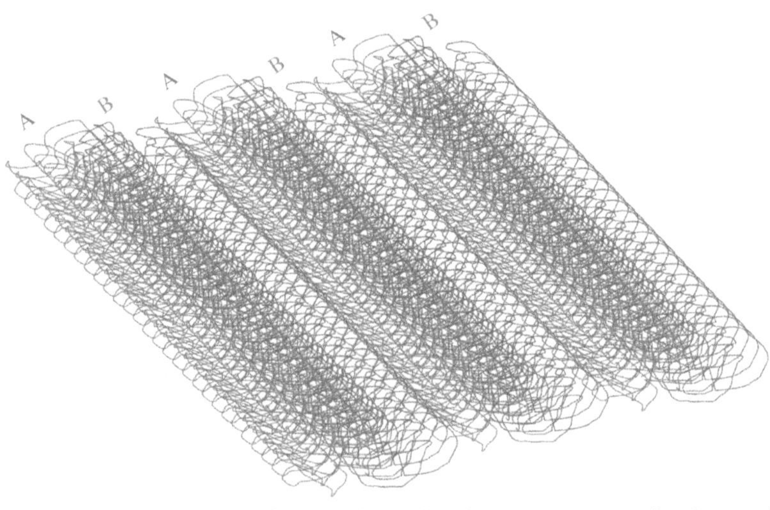

Figure 4.7 Schematic of a self-assembly, ordered nanostructure of a binary block copolymer, the dimensions of which are defined by the conformation of the two polymer blocks A and B.

thin film. The space between each phase is mainly decided by the molecular weight or repeat units of each block chain length. Therefore, controlling molecular weight is one effective way to control template dimension.

Block copolymer materials are similar to low-molecular weight non-ionic surfactant solutions (to be discussed in the following section) with respect to their general phase behaviour. Combination of inorganic siliceous components in a hybrid material with block copolymers is appealing for various reasons. The Wiesner[17] research group has recently developed nanostructured hybrid materials prepared by employing block copolymers as structure-directing agents. Block copolymer self-assembly has shown advantages to structure-direct materials with compositional and morphological control at the nanometre scale. Nanostructured porous ceramics, ceramic-carbon composites and ceramic-carbon-metal composites can all be developed based on the polymer templates.

A block copolymer has desirable macroscopic properties with predicted or known chemical, physical, thermal and mechanical properties of the final product. The architecture, chain length and composition of the block copolymer can be varied, which makes it possible to fine tune the properties of the polymer template and the corresponding inorganic materials developed from it. It is also known that the nature of the microstructure of block copolymers is determined by the characteristic length of the polymer chains and their corresponding conformation, which typically ranges from 5 to 100 nm. Guided by thermodynamic principles as discussed in this section, block copolymers can make many mesoporous materials with compositional and morphological control. Both small and large pore sizes as defined by sizes of different block chain conformations can be made with the aid of theoretical prediction.

The Wiesner group[17] investigated a sol-gel process of a mixture of two metal alkoxides, (3-glycidyloxypropyl)trimethoxysilane, $(CH_3O)_3Si(CH_2)_3OCH_2(CHCH_2—O)$ (GLYMO), and aluminium tri-sec-butoxide, $Al(O^sBu)_3$, with poly(isoprene-b-ethyleneoxide) block copolymers (PI-b-PEO) (Figure 4.8). This block copolymer has two important features. Firstly the products of metal alkoxide hydrolysis will preferentially swell the hydrophilic PEO block as they can act as solvents and their solubility parameters are much closer to PEO than PI. For example, the solubility parameter of ethyl alcohol CH_3CH_2OH is 11.27 $(\text{cal cm}^{-3})^{0.5}$ while the solubility parameters for PEO and PI are 9.9 and 8.1 $(\text{cal cm}^{-3})^{0.5}$, respectively. The Flory-Huggins interaction parameter is estimated to be 0.136, calculated by Equation (4.17), which gives the minimum degree of polymerization $N_c = 20$ meaning the

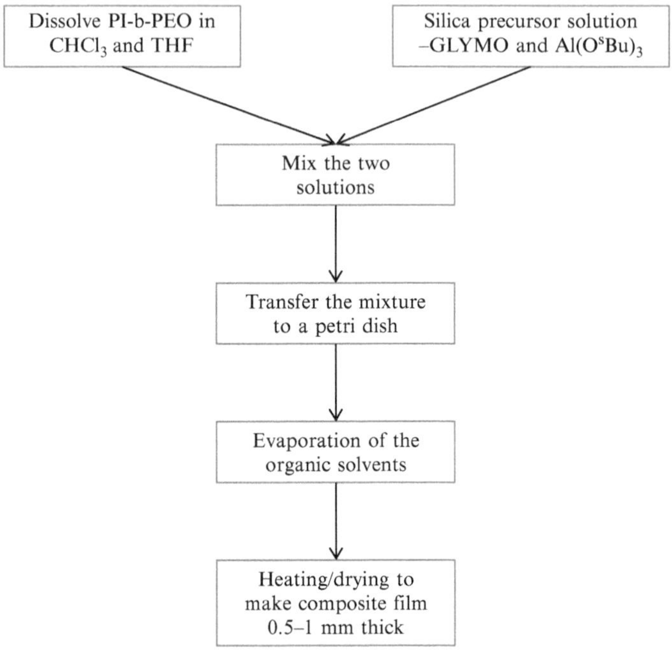

Figure 4.8 Process flowchart for the preparation of mesoporous silica.

copolymer used will display phase separation because the degree of polymerization is greater than 20. Second, the low glass transition temperature $T_g \sim 213$ K of the PI block introduces high mobility at ambient temperatures and should allow rapid formation of structures with long-range order. Figure 4.9 shows the nanostructured silica developed using a PI-PEO block polymer template. It is seen that a well regulated and ordered nanostructure of silica was developed, which is shown in the bright phase in the SEM where silica sits in the template of a size in the order of 10 nm.

More recently, nanostructured silica with various porous structures and morphologies were developed by using copolymers. Amphiphilic block copolymers, such as binary block polystyrene-block-poly(ethylene oxide) (PS-b-PEO) and triblock poly(ethylene oxide)-block-poly(propylene oxide)-block-poly(ethylene oxide) (PEO-b-PPO-b-PEO), have been used as soft structure-directing agents in the synthesis of nanostructured porous inorganic solids with controlled porous structures and morphologies.[18] The processing methods for organizing materials are the key. Evaporation-induced self-assembly is one of the simplest processes that enables the rapid production

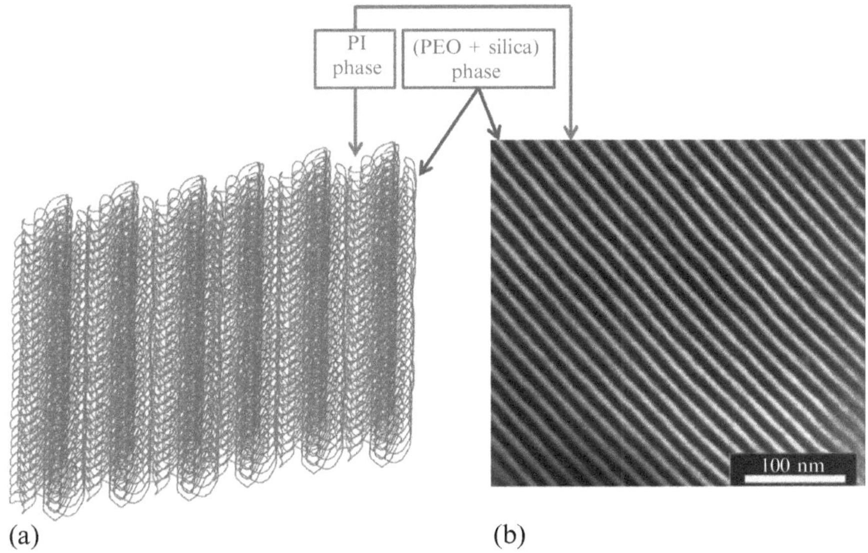

Figure 4.9 Silica embedded in an ordered nanotemplate formed by a PI-PEO block polymer: (a) model and (b) SEM where the brighter phase is silica embedded in PEO. *(SEM reprinted from Ref. 17 with Permission from AAAS).*

of patterned porous or nanocomposite materials. Figure 4.10 is a summary of the process, and Figure 4.11 shows the mesoporous silica made using this simple technology.

General processing of solvent evaporation technology includes the following:

Step 1: prepare stock solutions: (A) a solution containing block copolymer and (B) silica precursor (such as TEOS).

Step 2: mixing stock solutions A and B.

Step 3: casting/impregnation into a micromoulding with required geometry of the final 'products' (such as different shapes in form of fibres, discs, cylinders, etc.). The micromoulding has internal structures at nano-, micro- up to millimetre scales.

Step 4: drying of the cast products held in the micromoulding.

Step 5: removing the polymer templates and micromoulding in one go or step by step by employing suitable technology (such as calcinations, solvent/s).

Step 6: further treatment to get the desired mesoporous silica product.

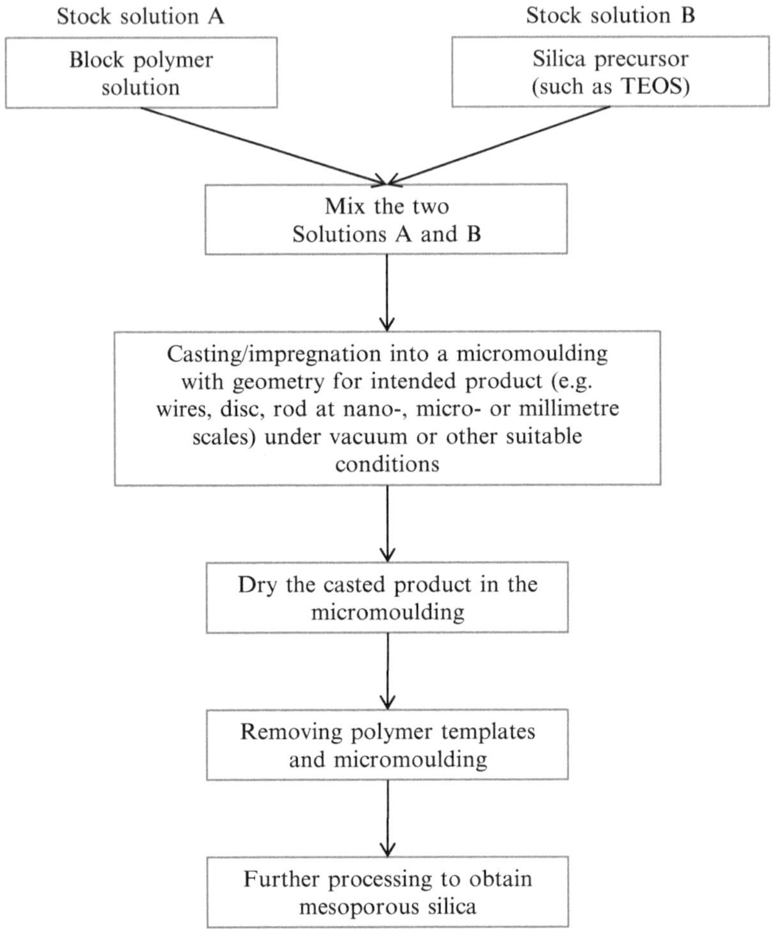

Figure 4.10 Process flowchart for the preparation of mesoporous silica structures by solvent evaporation.

4.3.2 Ordered Mesoporous Silica *via* Emulsion Chemistry

The concept of low-molecular-weight surfactants for self-assembly was first studied in the 1990s (Tables 4.1 and 4.2).[7,19–22] Ordered nanostructure was developed amongst which the preparation of mesoporous silica by sol-gel technology was the most intensively researched area. The ordered nanostructured silica is synthesized *via* sol-gel polymerization reactions in the presence of surfactants that act as pore-forming templates. The sol-gel process was initiated from different sources of silica including sodium silicate and silicon alkoxides such as TEOS and TMOS. The self-assembly of the

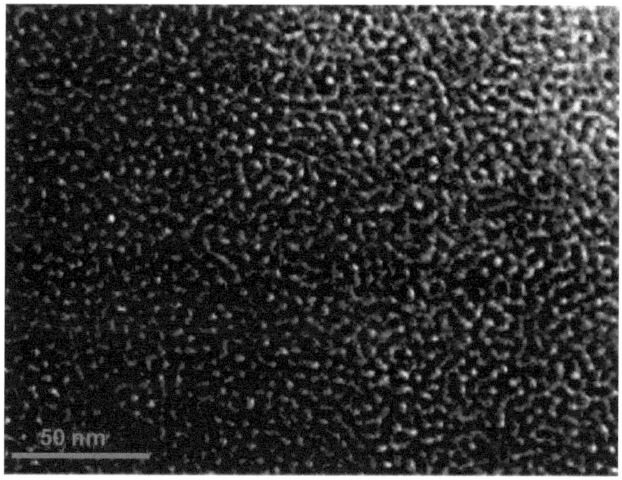

Figure 4.11 Microscopy image demonstrating the solvent evaporation approach for development of silica mesostructures.

Table 4.1 Common hydrophilic groups found in commercially available surfactants

Class	Structure
Polyethyleneoxide (PEO)	$R—OCH_2CH_2(OCH_2CH_2)_nOH$
Polyols	Sucrose, sorbitan, glycerol, ethyleneglycol, etc.
Polypeptide	R—NH—CHR—CO—NH—CHR′—CO—...—CO_2H
Polyglycidyl	$R—(OCH_2CH[CH_2OH]CH_2)n—...—OCH_2CH[CH_2OH]CH_2OH$
Sulphonate	$SO_3^-M^+$
Sulphate	$ROSO_3^-M^+$
Carboxylate	$R—COO^-M^+$
Phosphate	$R—OPO_3^-M^+$
Ammonium	$R_xH_yN^+X^-$ (x=1-3, y=4-x)
Quaternary ammonium	$R_4N^+X^-$
Betaines	$RN^+(CH_3)_2CH_2COO^-$

Data source: www.chm.bris.ac.uk.

surfactants produced various nanostructured hybrids. By removing the surfactant/s, silica materials with highly ordered nanostructure were obtained with pore sizes up to 10 nm. Tables 4.1 and 4.2 list the most commonly used, commercially available surfactants, all of which are potential candidates to act as templates for the development of mesoporous silica.

Table 4.2 Common hydrophobic groups used in commercially available surfactants

Group = R	Structure
Fluorocarbons	$CF_3(CF_2)_nCOOH$ (n = 4-8 linear or branched)
Silicones	$CH_3O[Si(CH_3)_2O]_nCH_3$ (n = degree oligomerization)
Natural fatty acids	$CH_3(CH_2)_n$ (n = 12-18)
Olefins	$CH_3(CH_2)_nCH{=}CH_2$ (n = 7-17)
Alkylbenzenes	$CH_3(CH_2)_nCH_2Ph$ (n = 6-10 linear or branched)
Alkylphenols	$CH_3(CH_2)_nCH_2PhOH$ (n = 6-10 linear or branched)
Polyoxypropylene	$CH_3CH_2(X)O(CH(CH3)CH_2)_nH$ (n = degree oligomerization; X = oligomerization initiator

Data source: www.chm.bris.ac.uk.

4.3.2.1 Non-Ionic Surfactant Templates

The most commonly used non-ionic surfactants are alkyl, fluorinated and silicone-based PEO surfactants:

$$CH_3(CH_2)_m-(OCH_2CH_2)_nOH,$$

where m = 8-20 and n = 4-30

$$CF_3(CF_2)_m-(OCH_2CH_2)_nOH,$$

where m = 4-8 and n = 10-30

$$CH_3O\left[Si(CH_3)_2O\right]_m-(OCH_2CH_2)_nOH,$$

where $m > 10$ and n = 10-30.

Low-molecular weight non-ionic surfactants can be treated in a similar way as binary block copolymers. They have much shorter chains than polymers, though the two parts of the surfactants have very different properties—one part is hydrophilic and the other extremely hydrophobic. Basic thermodynamic theories discussed in previous sections are still applicable to the non-ionic surfactants. The phase diagrams discussed in Figure 4.6 are also valid when non-ionic surfactants are used as templates. This means that a careful choice of surfactant can lead to the formation of spheres, cylinders, gyroids or lamellae templates as designed and as was seen with block copolymers. Based on this, various nanostructured silicas and other types of inorganic materials can be developed.

Zhao[23] has successfully developed a family of highly ordered mesoporous (2-30 nm) silica structures. They used commercially available non-ionic surfactants including PEO, oligomeric surfactants, and poly(alkylene oxide) block copolymers. Periodic arrangements of mescoscopically ordered pores based upon cubic, hexagonal, and lamellar templates were successfully made. Surfactants with short EO segments tend to form lamellar nanostructured silica at room temperature. Hexagonal nanostructured silica is favoured at higher temperatures, in the region of 100 °C. Highly ordered hexagonal mesoporous silica structures (SBA-15) with unusually large $d(1\,0\,0)$ spacings of 10-32 nm have been synthesized in the presence of triblock poly(ethylene oxide)-poly(propylene oxide)-poly(ethylene oxide) (PEO_{20}-PPO_{70}-PEO_{20}) copolymers.[24] The SBA-15 has a surface area of 690-1040 $m^2\ g^{-1}$ (BET), pore sizes of 4.6-30 nm, wall thicknesses of 3-6 nm, and pore volumes as large as 2.5 $cm^3\ g^{-1}$.

Bagshaw[21] and Smarsly[25] employed PEO as a template-forming surfactant. Hydrogen bonding interactions between the hydrophilic surfaces of flexible rod- or worm-like micelles and $Si(OEt)_{4-x}(OH)_x$ hydrolysis products results in the self-assembly of an inorganic oxide framework. Figure 4.12 shows uniform diameters ranging from 2.0 to 5.8 nm were obtained by varying the size and structure of the surfactant molecules.

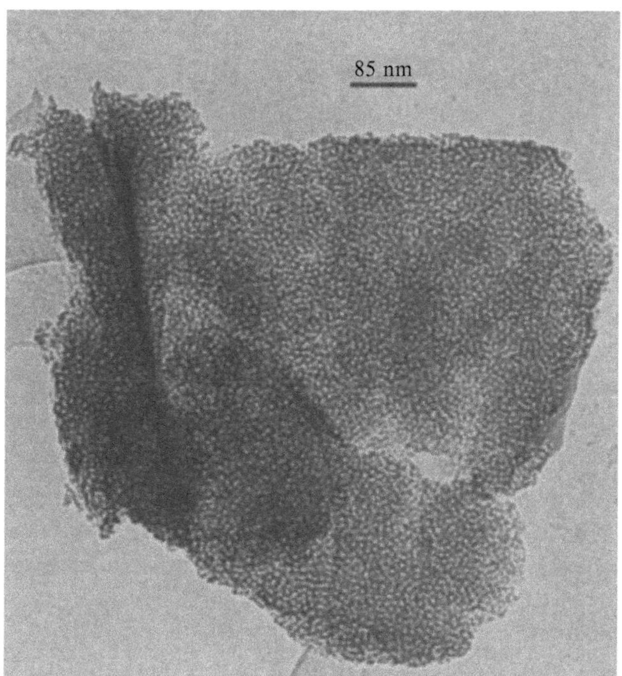

Figure 4.12 Microscopy results of approach for development of silica mesostructures. *(Reprinted with permission from Ref. 25. Copyright 2001 American Chemical Society).*

Table 4.3 Theoretical parameters and experimental pore sizes of mesoporous silica

Non-ionic surfactant	Molecular weight $(CH_2CH_2)_n$	Molecular weight $(CH_2CH_2O)_m$	Average density (g cm^{-3})	$N^*\chi_{AB}$	$r_{distance}$ (nm)	Pore size (nm)
C8E4	97	257	1.05	3.95	1.85	2.22
C12E4	145	257	1.02	5.72	2.07	2.66
C12E8	145	513	1.06	12.63	2.43	2.71
C16E8	193	513	1.05	15.70	2.61	3.05
C12E30	145	1921	1.10	98.18	3.58	3.31
C16E10	193	641	1.056	20.64	2.76	3.51
C18E20	217	1281	1.08	58.42	3.41	3.77

With reference to Smarsly's work,[25] the non-ionic surfactants —$(CH_2CH_2)_m$—$(OCH_2CH_2)_n$— and the corresponding mesoporous dimensions formed by the surfactant can be predicted. Table 4.3 highlights some theoretical parameters and the corresponding experimental results of expected pore sizes of some mesoporous silicas. The surfactant Flory-Huggins interaction parameter $N^*\chi_{AB}$ varies as calculated according to Equation (4.17). The smallest value is 3.95 and the largest 98.18. Therefore, the two parts of the non-ionic surfactant (i.e., —$(CH_2CH_2)_m$— and —$(OCH_2CH_2)_n$—) are predicted to adopt a state of phase separation when dried after formation of the silica network. The nanopore size varies too, generally proportionally to the conformational size of the surfactants represented by the dimension under the undisturbed states. Figure 4.13 is a plot showing the relationship between theoretical prediction and the actual measured pore size of a range of mesoporous silicas. There is good agreement between the two; the predicted values are close to the experimental results. However, the predicted values tend to be smaller than the experimentally determined values. With further development of the theories discussed in the previous section, it is expected that the pore size of mesoporous silica can be more accurately predicted and aligned to the experimentally found values, the significance of which is not only in theoretical determination, but in the design and development of ordered nanostructured silica and other inorganic materials.

Notes for Table 4.3: all parameters calculated are based on the raw data in the reference 'Polymer Handbook' 3rd edition.[12] (i) Average densities were calculated based on the amorphous densities of PE (polyethylene), 0.85 g cm^{-3} and PEO, 1.123 g cm^{-3}; (ii) χ_{AB} was calculated according to Equation (4.16) then $N\chi_{AB}$ was calculated using values for δ_{PE} and δ_{PEO} of 8.0 and 9.9 (cal cm^{-3})$^{1/2}$; (iii) the undisturbed chain dimension represented by surfactant chain end to end distance $r_{distance}$ was calculated.

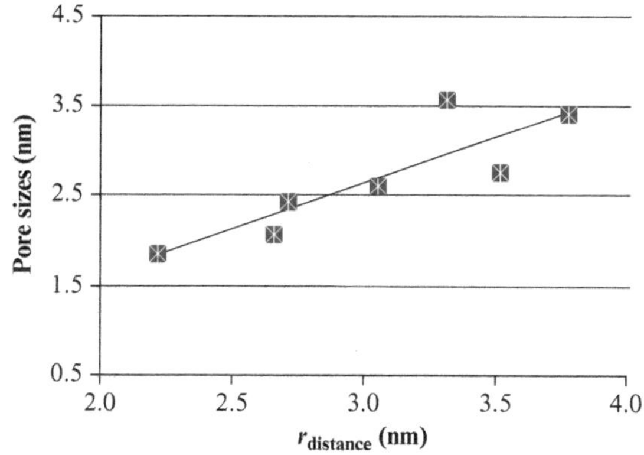

Figure 4.13 Relationship between theoretical prediction and measured pore size of mesoporous silica. *(Data: Table 4.3).*

4.3.2.2 Ionic Surfactant Templates

The use of ionic surfactants to develop mesoporous templates was a ground breaking process first reported in the 1990s for the synthesis of ordered inorganic nanostructured materials. Synthesis conditions such as source of silica, type of surfactant, ionic strength, pH and composition of the reaction mixture, temperature, and duration of synthesis affect the ionic surfactant micelle conformation, the silica-surfactant interactions and the degree of silica polymerization. The well-known mesoporous M41S materials were made using ionic surfactant template technology. The pore diameters made *via* this approach are ~2-10 nm with amorphous pore walls. The most well-known representatives of this class include the silica solids MCM-41 with a hexagonal arrangement, MCM-48 with a cubic arrangement, and MCM-50 with a laminar structure,[8,21,22] which are shown in Figure 4.14.

Recall the phase diagram (Figure 4.6 from Section 4.3.1.10) discussed with regard to binary block copolymers. Increasing the interaction parameter $N\chi_{AB}$ and composition ϕ_A from 0 to 1 will lead to the formation of templates starting from sphere, cylinder, gyroid, lamellar then, at $\phi_A > 0.5$, gyroid, cylinder, and back to sphere again. It is expected that the ionic surfactants will display similar phase behaviour to binary block copolymers. One block of the polymer chain can be equivalent to the hydrophobic portion of the ionic surfactant, whilst the other block, if we were to reduce its size equivalence to that of a monomer, can be equivalent to the cationic or anionic charged portion of the surfactant molecule.

Figure 4.15 is a plot that demonstrates the typical processing and reaction mechanisms. The surfactant molecules contain short hydrophobic chains but have strongly hydrophilic functional groups at one end.

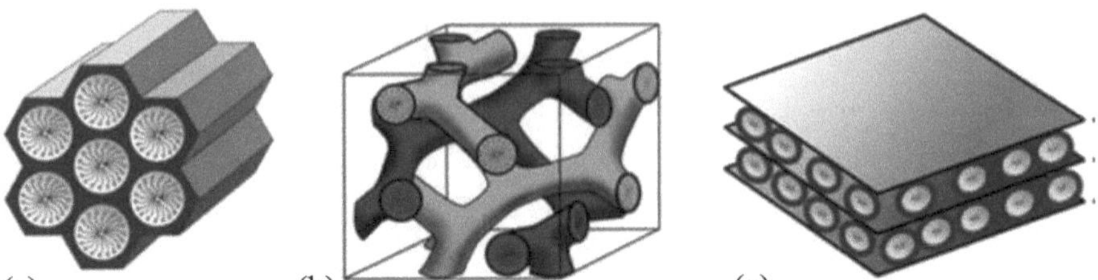

Figure 4.14 Structures of mesoporous M41S materials: (a) MCM-41 (2D hexagonal), (b) MCM-48 (cubic) and (c) MCM-50 (lamellar). *(Reprinted with permission from Ref. 8. Copyright 1992 American Chemical Society).*

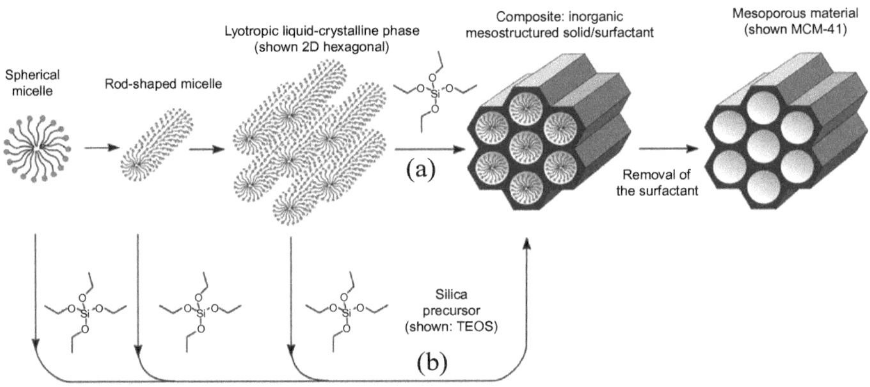

Figure 4.15 Formation of mesoporous materials using surfactants to form mesoporous templates: (a) true liquid-crystal template mechanism and (b) cooperative liquid-crystal template mechanism. *(Reprinted with permission from Ref. 8. Copyright 1992 American Chemical Society).*

Following the same logic of the phase diagram shown in Figure 4.6 of a binary block copolymer system, the ionic surfactant system will form spheres at lower concentrations, then move to cylinder, then gyroid up to lamella structure with increasing surfactant concentration. These structures are the templates upon which ordered nanostructured silica composites are based and subsequently formed through sol-gel processing. The mesoporous silica is obtained by subsequent removal of the surfactant by solvent-based extraction or calcination.

Figure 4.16 highlights the use of mesoporous silica MCM-41 for the controlled release of ibuprofen.[27] The mesoporous silica was impregnated with ibuprofen up to 34% (weight) and the dissolution study carried out in simulated body fluid (SBF) at 37 °C. Without employing controlled release technology, the ibuprofen release reaches 100% very quickly under

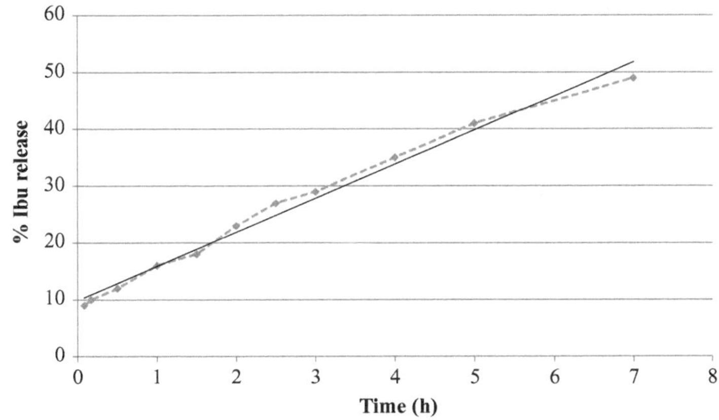

Figure 4.16 Controlled release of ibuprofen from mesoporous silica MCM-41. *(Adapted from Ref. 27 Copyright © 2006 Elsevier Masson SAS. All rights reserved).*

the same conditions. This ibuprofen-loaded silica MCM-41 demonstrates the effectiveness of the inorganic CRT. It can also be seen that diffusion is the release controlling mechanism up to ~45% of ibuprofen release as indicated by the solid trendline shown in Figure 4.16.

Figure 4.17 demonstrates the controlled release of atenolol from SBA-15.[28] To evaluate the performance of SBA-15 mesoporous silica for drug controlled release, silica discs were soaked in a saturated solution of atenolol and the delivery solvent subsequently removed. The release profile was obtained in SBF at room temperature. UV spectrophotometry was used to monitor the amount of drug delivered as a function of time. It is confirmed again that diffusion is the release controlling mechanism up to ~43% of atenolol release, as indicated by the dotted straight line shown in Figure 4.17.

Despite successes in the synthesis of mesoporous materials in various meso- and macroscopic forms, the pore size of nanostructured materials obtained from ionic surfactants is in general limited. This limitation is caused by the surfactant size and the restricted range of compositions. On the other hand, block copolymers have shown more promise in overcoming these two drawbacks. The use of block copolymers should have great potential for further future development because there are so many design parameters to choose from and vary. Further research should result in achieving better control over the formation of the desired mesoporous structures for a wide range of applications. The fundamental understanding of the thermodynamics of the block copolymer systems is also essential in order to have a good

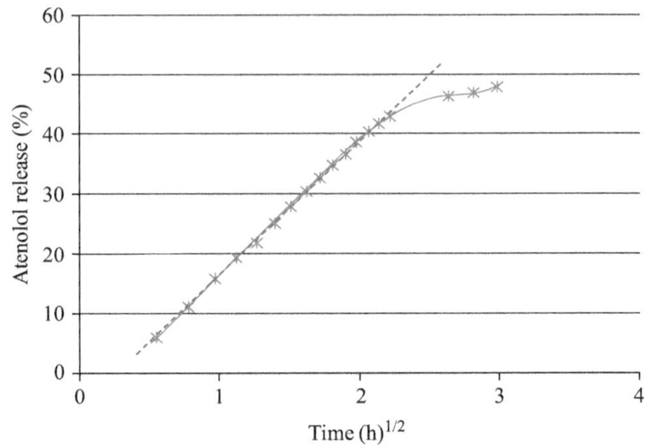

Figure 4.17 Mesoporous silica SBA-15 as a controlled release delivery system for atenolol. *(Reprinted from Ref. 28 under a Creative Commons Attribution Licence; http://creativecommons.org/licenses/by/4.0/).*

compositional and morphological control of ordered mesoporous functional materials for better design and development.

One of the drawbacks to the preparation of ordered nanoporous silica molecules, using ionic or non-ionic templates, is that further processing is required post-formation to remove the template molecules. As such the isolation of pure silicate-based carriers becomes more difficult and drawn-out. However, more recent technological advances and improvements in processing methodology are now beginning to overcome some of these issues, and a new extended branch of mesoporous silicate production will be the focus of the following section. Although this field of materials research and nanotechnology is still very much in its infancy, there is great potential in the use of ordered mesoporous silica materials for a potentially huge range of inorganic CRT applications.

4.4 NON-ORDERED POROUS SILICA FOR CONTROLLED RELEASE

As described in the previous section, organic sacrificial pore-forming agents are used to create porous silicates with both homogeneous pore sizes and pore size distributions. A prior knowledge of the typical sizes of the micelle structures formed in solution, during the preparation of the polysilicate network, can lead to the controlled formation of a porous matrix with porosity in the micro-, meso-, and macro-porous ranges as desired. This structure-directed pore-forming methodology is by far the most prevalent method being used by research groups active in this area and by the small number of commercial entities attempting to bring such products to market. The technology for making such materials is relatively modern,[8,19,21,22] whilst the preparation of non-ordered porous silicates was demonstrated well before the first published synthesis of ordered porous silica. The first reported silica gels were made in 1846 by M. Ebelmen at the 'Manufacture de Ceramiques de Sevres' in France.[2,3] This methodology began to become interesting from an industrial perspective in the late sixties and seventies when the controlled hydrolysis and condensation of alkoxides for the preparation of multi-component glasses was successfully developed.[29,30] The modern development of sol-gel science and technology did not start until the first 'International Conference on Glasses and Glass Ceramics Obtained from Gels' in 1981. During the early stages of the technological development process, researchers were not overly concerned on controlling the porosity of the product materials, but more on whether crack-free monoliths could be prepared. Subsequently though, the development and control of the nanoporous structures of the product glasses was found to be very important for a number of potential applications and underpins the area of inorganic CRT science and technology discussed in this text.

Moving away slightly from the hydrolysis and condensation reactions of silicon alkoxide precursors, of significant industrial importance is the preparation of porous silica materials by the precipitation reactions of sodium silicate. Sodium silicate species can be extracted from naturally occurring sources such as diatomaceous earth (DE) and rice husk and are then neutralized with mineral acid. Synthesizing porous silicates in this way does not require the use of pore-forming agents and by definition removes the need for eliminating these sacrificial agents. This much simpler processing route is what makes it such a popular process, particularly on the industrial scale. These silica materials though have found only limited use as controlled release drug delivery systems since it is difficult to control the release dictating properties during the synthetic process.

Of greater importance to the research community, as evidenced by the huge increase in publications in this area, are the hydrolysis and condensation reactions, more commonly referred to as the sol-gel reactions, of organic silicon alkoxides. A great deal of work has been carried out into how varying the processing parameters of the sol-gel reactions of silicon alkoxides can have dramatic effects on the physical properties of the product porous materials, and as a result, reasonably tight control over the release of any biologically important encapsulated molecules can be achieved. Such promise and versatility in these materials as controlled release drug delivery systems has seen them applied across a broad range of therapeutic areas and administration methods.

4.4.1 Porous Silica by Precipitation Chemistry

As an alternative to the sol-gel reactions of organic silicon alkoxides, porous silica can be formed by the acid-base chemistry of solutions of sodium silicate. This form of porous silica, also commonly referred to as precipitated silica, typically has an average pore size of 30 nm and a BET specific surface area of 5-100 $m^2\ g^{-1}$ (refer to Chapter 3 for material characterization). During a typical precipitation process, as shown in Figure 4.18, sulphuric acid and sodium silicate are continuously added to a mixture of water already containing a little sodium silicate. The presence of the sodium silicate in solution to begin with helps to buffer the initial addition of the acid to the reaction mixture.

Figure 4.19 shows the change in particle size and morphology as the precipitation process occurs. The condensation reactions of silicic acid are pH dependent and in the presence of an acid catalyst, monomers condense to form oligomeric structures which are kept in solution *via* solvation. Eventually these primary solvated molecules become large enough to form solid particles and a colloidal suspension is formed. These colloidal particles continue to grow by monomer addition leading to the formation of nanoparticles. However, the simultaneous addition of sodium ions results in retardation of particle growth due to the extra ions present masking the electrostatic-driven bond-forming reactions and flocculation of the particles occurs in preference to particle growth. The point at which particle

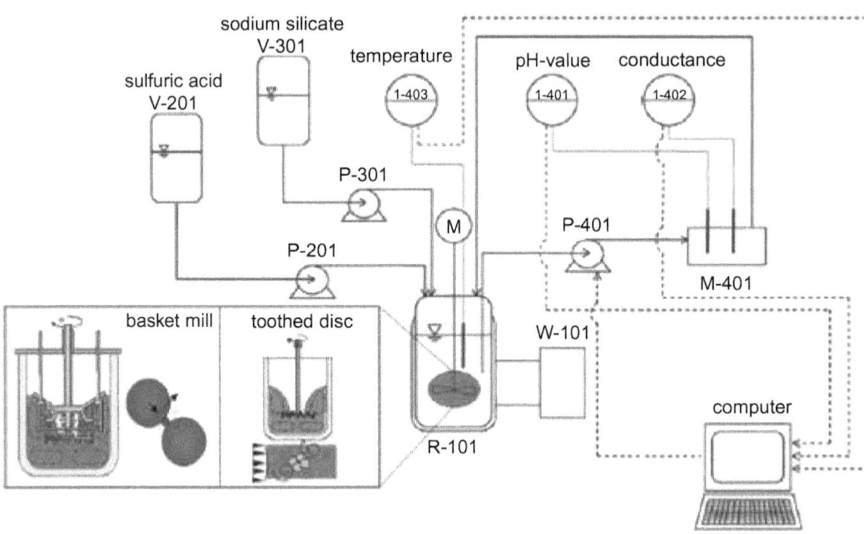

Figure 4.18 Schematic representation of the silica precipitation experimental set-up. *(Reprinted from Ref. 31 with permission from Elsevier).*

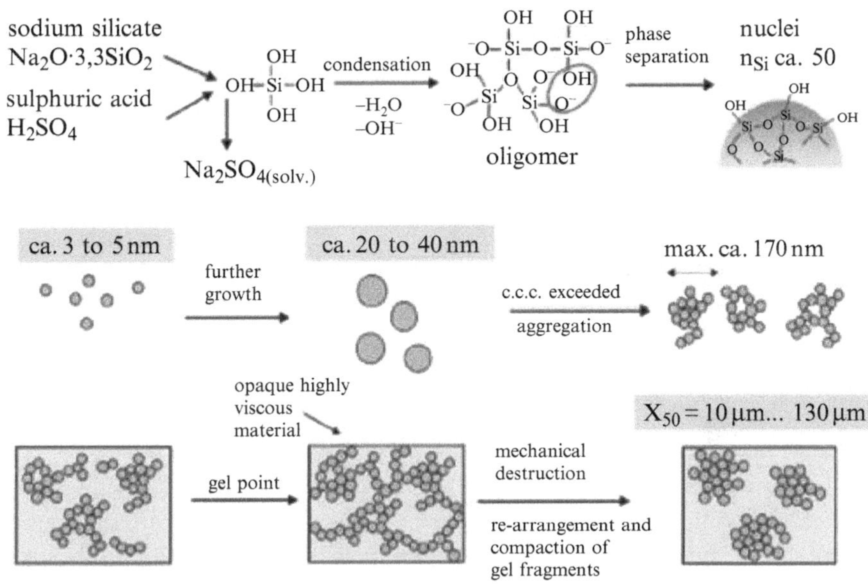

Figure 4.19 Schematic diagram of the silica precipitation process. *(Reprinted from Ref. 32 with permission from Elsevier).*

aggregation occurs preferentially is dependent on a number of factors including the size of the particles, the size of the critical coagulation concentration, the particle charge and the particle concentration. The viscosity of the colloidal suspension increases as the particle concentration increases until a gel is formed. A method of processing the gel phase further is by consolidation of the gel network by mechanically breaking it up. The product particles are then stabilized by bringing the pH to the isoelectric point of silica.[32]

In an effort to conform to a more green economy, and more effectively recycle and reuse natural resources, the important silicate precursor molecules are now being sourced from alternative sources such as rice husk ash, a by-product of rice husk pyrolysis. Extraction of silicate salts is carried out using boiling sodium hydroxide, following which a silica hydrosol is formed by neutralizing the solution using sulphuric acid. Silica gels with porosity ranging from 9% to 85% can then be obtained *via* modification of the ageing and drying processes.

As an extension of this basic processing methodology, the addition of 10% (by volume) of TEOS, followed by ageing the resulting gel in ethanol, results in a silica product that is highly porous (pore volume: 0.78 $cm^3\ g^{-1}$). Exclusion of TEOS and ageing of the gel in water, as opposed to ethanol, both independently result in less porous products (see Table 4.4).[26]

Table 4.4 Physical properties of the silica gels, dried and aged by different methods

Sample	Density (g cm^{-3})	Porosity (%)
Silica gel with TEOS—ethanol as pore liquid	0.32	85.0
Silica gel without TEOS—ethanol as pore liquid	0.66	69.8
Silica gel without TEOS—water as pore liquid	1.00	54.3

Adapted from Ref. 26.

When the liquid phase dispersed within the gel structure is water, removal of the solvent by drying/evaporation results in large capillary forces being exerted on the remaining silica structure pore walls. As a result, the microstructure of the silicate product shrinks and the pore volume and average pore size decrease. By replacing the water with ethanol prior to drying of the silica gel products, the capillary forces exerted on the remaining silicate structure during the drying process are reduced, resulting in decreased shrinkage of the pore cavities during removal of the ethanol. This results in an increase in product porosity (54.3-69.8%) and a decrease in material density (1.00-0.66 g cm^{-3}). The addition of TEOS during the silicate network-forming process acts to increase the extent of cross-linking of the network, decreasing the shrinkage of the pore cavities further during the drying process. As a result, product porosity is increased further (69.8-85.0%) and product density is decreased (0.66-0.32 g cm^{-3}).

Advantage is also being taken of nature's ability to generate porous inorganic systems. Biologically generated three-dimensional porous structures are formed by diatoms, single-cell algae. Diatoms themselves are freely available by cultivation but are more cheaply sourced from DE, which is readily available as a by-product of the mining industry. The 3D porous networks present in diatomaceous earth have a pill-box type structure containing levels of both micro- and mesoporosity (Figure 4.20).

In order to make extracted DE pure enough for pharmacopoeial applications, it can be first added to a polar protic solvent such as water and subsequently sieved to remove any water-soluble impurities. The retained solids are then subject to conditions of boiling mineral acids such as hydrochloric, sulphuric or nitric acid and again residual solids are collected. These are then heat treated to 225-275 °C in a furnace. The product solid is then treated with an oxidizing agent such as dilute hydrogen peroxide and the solid remaining from this process is subjected to a final drying process. The advantage of purifying the DE in this manner is that the natural microstructure of the DE can be retained such that its desirable physical properties, primarily the porous structure in the current context, can be preserved.[34]

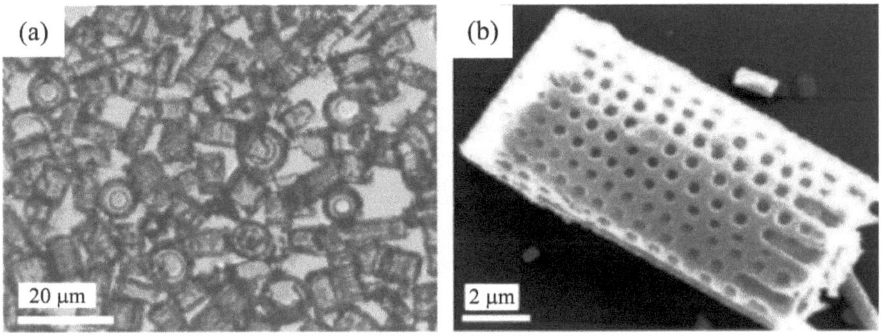

Figure 4.20 SEM images of pure diatoms obtained from diatomaceous earth. *(Reprinted from Ref. 33 with permission from Elsevier).*

The use of naturally sourced porous diatoms to deliver the poorly water-soluble, non-steroidal anti-inflammatory compound, indomethacin was explored. Wetting of the silica surface with ethanol was first carried out under conditions of sonication. An indomethacin solution in ethanol was then added to the wetted DE, and drug-loaded DE was obtained following evaporation of the residual ethanol (Figure 4.21).

The release kinetics were evaluated following both an oral delivery dissolution model and an implant delivery dissolution model. Oral delivery was modelled by carrying out dissolution testing in pH 7.2 phosphate buffer under sink conditions. Complete release of indomethacin was noted at 6 hours, an extension on the 3 hours seen with pure indomethacin.

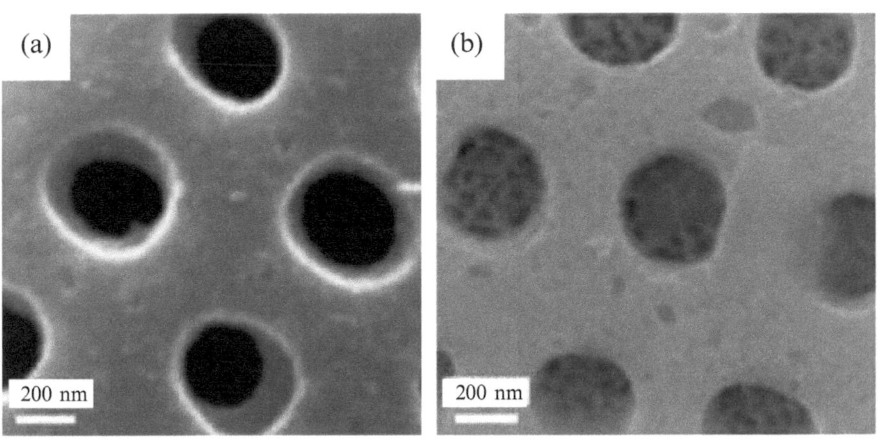

Figure 4.21 High-resolution SEM images of diatom pores before and after indomethacin loading. *(Reprinted from Ref. 33 with permission from Elsevier).*

Modelling for an implant delivery system was carried out under non-sink conditions. A distinctly two-phase type of release was observed. A relatively large burst release of 65-70% was first seen during 6 hours, followed by a zero-order type release over the next 12 days. It is suggested that the burst release is a function of the diffusion-controlled release of indomethacin adsorbed to the outer surface of the DE microcarriers. Whereas the subsequent zero-order release is dictated by release of indomethacin encapsulated within the porous structure of the diatomaceous earth (Figure 4.22).[33]

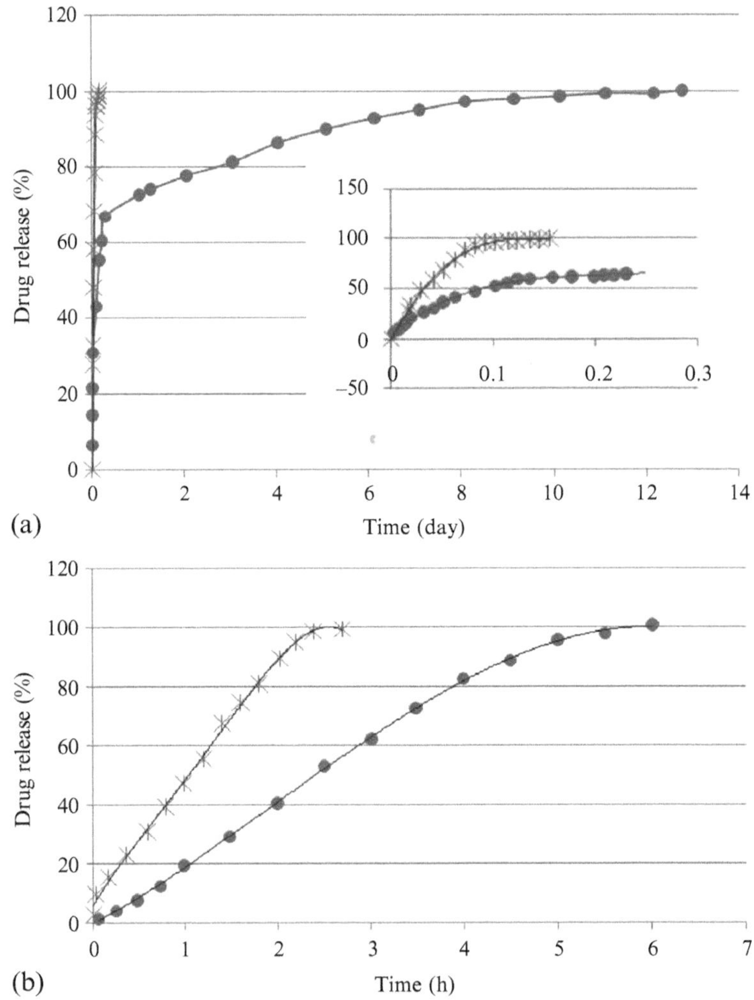

Figure 4.22 Drug release curves for indomethacin-loaded diatom microshells. (a) Dissolution under non-sink conditions (an implant model) and (b) dissolution under sink conditions (an oral formulation model). *(Adapted from Ref. 33 with permission from Elsevier).*

This study demonstrated the successful use of naturally sourced porous silica for the encapsulation and delivery of the poorly soluble compound indomethacin. Indomethacin could be loaded up to 22% by weight and displayed two very distinct phases of release—a first-order release over the first 6 hours followed by a zero-order release over the course of 2 weeks.

4.4.2 Non-Ordered Silica by the Sol-Gel Chemistry of Silicon Alkoxides

The preparation of ordered mesoporous silica materials is by far the most widely researched methodology for preparing silica-based inorganic materials with nanofeatures. The use of polymer-templating during the sol-gel synthesis of ordered mesoporous silica leads to products with a uniform and narrow pore size distribution. In addition, the use of emulsion chemistry can lead to porous particles with regular particle size—generally spherical in nature. In combination, these two methodologies can lead to hollow mesoporous silica nanoparticles. However, in order to remove the pore-forming template molecules the complexes must be thermally treated, which can compromise the structural integrity of the porous products. Alternatively the templating molecules can be removed by dissolution in a suitable solvent system, though this can be a lengthy process which may still not completely remove all the pore-forming molecules.

Prior to the development of ordered mesoporous silica materials, the production of non-ordered porous silicas was well known. Silica gels or xerogels formed by aqueous sol-gel reactions are characterized by their non-uniform pore structures; having disordered pore structures with wide pore size distributions. Their preparation is simpler as no polymer templates are required, the pore-forming molecules are the solvents of the reaction process; water, methanol, ethanol, and the by-products formed from the hydrolysis of the silicon alkoxide precursors. As such, they are typically microporous (<2 nm) or mesoporous (2-50 nm) in nature (IUPAC Gold Book definition).

Despite the synthesis of non-porous silica materials being well-known historically, it was only with the re-emergence of research into porous silica following the first preparations of ordered mesoporous silicas that the potential of non-ordered porous silica for controlled applications was realized. Key to the performance of this sub-set of porous silica materials is controlling not only the pore size, but also the pore size distribution.

4.4.3 Processing of Non-Ordered Porous Silica

The potential for silica-based porous inorganic materials for drug delivery was first realized in 1972.[35] Fumed silica was used as an insoluble excipient specifically for enhancing the dissolution kinetics of small molecule drug compounds with poor water solubility. The compounds indomethacin, aspirin, sulphaethidole, griseofulvin, reserpine, chloramphenicol, oxolinic acid, probucol and hydrochlorothiazide were adsorbed onto the silica surface and the effects of modification of the pH, the use of different wetting agents, and different levels of agitation intensity on the drug adsorption/desorption to the silica surface were investigated. In all cases an increase in the rate of dissolution from the silica carrier system as opposed to the freely crystalline drug was found. Fumed silica is a non-porous system so this increase in the rate of drug release is a consequence of increasing the contact area of the drug molecules to the dissolution medium and is purely a surface area effect.

For *in vitro* dissolution studies, as carried out here, drug diffusion through the solid-liquid interface surrounding the active formulation is the rate determining process. As such the rate of dissolution can be evaluated using the Noyes-Whitney equation.

$$\frac{\mathrm{d}W}{\mathrm{d}t} = \frac{\mathrm{DA}(\mathrm{C_s} - \mathrm{C_b})}{L} \tag{4.18}$$

where D is the diffusion coefficient, A the surface area, C_s the solubility of the drug, C_b the concentration of drug in the bulk solution and L is the thickness of the diffusion barrier.

For cases where *in vitro* dissolution is carried out under sink conditions, then $C_s \gg C_b$ and Equation (4.18) simplifies to

$$\frac{\mathrm{d}W}{\mathrm{d}t} = \frac{\mathrm{DAC_s}}{L} \tag{4.19}$$

From this expression it can be seen that any increase in the drug surface area is going to lead to an increase in the dissolution rate.

Similarly, Yang[36] investigated the adsorption of griseofulvin, prednisone, and digoxin onto amorphous silica and the effects on these compounds' dissolution rates. They found that adsorption of the drug molecules, again following a solvent deposition method, occurs within the porous structure of the silica material. As an alternative method for combining drug with silica carrier, they investigated co-ball-milling of the two substrates and found that the carrier system with the largest average particle diameter resulted in the slowest release.

Narurkar and Jarowski[37] followed this up with a similar study, investigating the effect of the drug-to-silica ratio on the controlled release of prednisolone. Again they investigated preparation of the formulations *via* solvent deposition and ball-milling in combination with the use of both non-porous and porous silicas. In a subsequent *in vivo* oral study with dogs, the efficiency of prednisolone absorption was independent of whether a porous or non-porous silica was used.

The above early examples of silica for drug delivery all feature the addition of the active to be delivered to a preformed carrier system. Braun was the first to add the active during the formation of the silica carrier; alkaline phosphatase was added to a polymerizing TMOS solution.[38]

Ellerby extended this idea to add functional biomolecules to the sol phase of a sol-gel process. Once permanently immobilized, the small molecule substrates for the enzymes copper-zinc superoxide dismutase, cytochrome *c*, and myoglobin were able to diffuse into the solvent-filled pores of the carrier matrix and their reactions monitored by visible absorbance spectroscopy.[39]

These early studies set the precedent for the addition of active molecules during the silica-forming process and from here on in this methodology has been extensively researched not only for the encapsulation of biomarker molecules, but also for the controlled release of biologically active agents.

As an extension of the work of Ellerby and Braun, Ducheyne was the first to successfully prepare an enzyme-loaded matrix for the purpose of demonstrating its sustained release.[40] By careful modification of the sol-gel-derived product silica, and a judicious choice of enzyme, he was successfully able to demonstrate the release of transforming growth factor-β1 (TGF-β1). The inclusion of protein-based biomolecules during the formation of the silica matrix is difficult since the acidic/basic conditions used to catalyse the hydrolysis and condensation network-forming reactions, and the common use of organic solvents, usually produce conditions capable of denaturing the *in situ* enzymes. Also, high temperatures are frequently used to further consolidate the glass network following its formation, which again are conditions to which proteins are incompatible. Ducheyne got around these restrictions by not adding any extra solvent to the reaction mixture, by allowing the final xerogel to form at room temperature, and by the use of the enzyme TGF-β1 which is highly stable under acidic conditions.[41]

Sieminska and Verda[42] were the first to include small drug molecules during the sol-gel processing of porous silica glass. They added the steroid compounds progesterone, estradiol, estrone and hydrocortisone to the sol-gel solution and compared the subsequent release of the small molecules from the glass carrier with that when these molecules had been loaded into a preformed porous carrier.

4.4.3.1 The Contribution of Matrix Dissolution to Drug Release

One of the advantages of adding the active to the glass-forming sol-gel process is that molecules that are larger than the typical nanosized porous features of the product glasses can still be encapsulated within the silica matrix. The inclusion of larger molecules such as viruses,[43,44] cells[45–48] and proteins[49–55] in this way results in a drug delivery system which is more heterogeneous in its microstructure, and as such, is more susceptible to hydrolytic cleavage. The relative contributions of active diffusion and matrix dissolution on the overall release kinetics from non-ordered silicas were investigated. Sol-gel process parameters were varied during the loading of the small drug molecule propranolol and the large molecule bovine serum albumin (BSA) and *in vitro* dissolution studies were carried out in two media—a tris(hydroxymethyl) aminomethane (tris) buffer and a SiO_2 saturated Tris buffer. The rationale being that in systems that require matrix dissolution for drug release to occur, no drug release will be seen since the presence of silica in the dissolution medium prevents the hydrolysis of the silica matrix.

The release of propranolol hydrochloride from silica xerogel monoliths showed similar dissolution rates into both SiO_2 saturated and unsaturated Tris buffers, indicating that diffusion is the primary drug releasing mechanism. In contrast, no BSA release was noted from silica xerogel monoliths in the saturated buffer whereas drug release was observed in the unsaturated medium where dissolution of the carrier matrix was able to occur.[56]

As evidenced by the above investigations, there are many different methods of altering the silica network-forming process in order to affect a change in the release kinetics of any incorporated active. In the design of such drug delivery systems, a clear idea of administration route and desired active delivery profile must be known. With these goals in mind, the key process parameters affecting the silica product pore structure need to be evaluated specifically with the active of choice. This methodology is slightly limited in that the active of choice needs to be soluble in the reaction sol-gel mixture, so that it can be found homogeneously distributed in the product xerogel. However, with such a wide range of processing parameters available to control the microstructure of the product silica carriers, there are many more options for fine-tuning the controlled release of your active of choice.

4.5 POROUS SILICA FOR IMPROVED EFFICACY AND TOLERABILITY OF TOPICALLY ADMINISTERED ACTIVE COMPOUNDS

The pipelines of many pharmaceutical companies, as well as the medicinal chemistry departments of many research organizations, are full of promising lead compounds and molecules that interact with their therapeutic receptors exceptionally well, but will never be suitable for administration because of their poor pharmacokinetic properties. The biggest downfall for most compounds in this category is their lack of aqueous solubility. The enhancement of pharmacokinetic properties for these types of compounds by encapsulating them within the porous structure of silica carriers is one of the most widely researched applications for these nanomaterials (Table 4.5).

However, a particular group of formulations employing porous silica for the controlled delivery of topically active pharmaceutical compounds is currently leading the way for mesoporous silica in terms of a commercial product coming to market.

Sol-Gel Technologies were founded in 1997 and have developed a platform for the delivery of a range of compounds for dermatological application, primarily as treatments for acne and rosacea.[76] An over-the-counter treatment for acne is pending commercialization, two further treatments for acne are in preclinical trials and a treatment for papulopustular rosacea has met its Phase II endpoints. The acne treatment, marketed as Cool Pearls™ Anti Acne Kit, and the rosacea treatment (SGT-VD-54) both feature benzoyl peroxide encapsulated within porous silica carrier molecules.

From these novel formulation platforms, benzoyl peroxide can be eluted at a rate of 20–80% per hour per weight of the originally encapsulated benzoyl peroxide. The researchers found that a release rate of less than 80% per hour improved the treatment of rosacea whilst reducing some of the more undesirable adverse effects. In a number of particular treatment pathways, the application of controlled release benzoyl peroxide was as effective as benzoyl peroxide used in conjunction with an antibacterial agent.

Table 4.5 Different mesoporous porous silicas used to enhance the dissolution kinetics of poorly water-soluble drug compounds

Poorly soluble drug	Therapeutic use	Porous silica carrier
Resveratrol	Anti-inflammatory	MCM-41, KIL-2[57,58]
Simvastatin	Treatment for hypercholesterolemia	Core-shell dual-mesoporous silica nanospheres[59]
Aceclofenac	NSAID	MCM-41[60]
Celecoxib	NSAID	Face-centred cubic mesoporous silica[61]
Fenofibrate	Treatment for hypercholesterolemia	SBA-15[62,63] MCM-41[63]
Carbamazepine	Anti-convulsant	Hollow mesoporous silica spheres[64] SBA-16[65]
Silybin	Anti-hepatotoxic, oncology	Porous silica nanoparticles[66]
Naproxen	NSAID	MCM-41[67]
Paclitaxel	Oncology	Mesoporous silica nanoparticles[68]
Silymarin	Anti-hepatotoxic	Porous silica nanoparticles[69]
Indomethacin	NSAID	SBA-16[70] Diatom microshells[33]
Carvedilol	Anti-hypertensive	SBA-16[71]
Itraconazole	Anti-fungal	Syloid-AL1, Syloid-244 (non-ordered mesoporous silica), thermally oxidized and thermally carbonated mesoporous silica[72] Ordered mesoporous silica[73]
Piroxicam	NSAID	SBA-15 occluded with Pluronic 123[74]
Telmisartan	Treatment for hypertension	Mesoporous silica nanoparticles[75]
Oxcarbazepine	Anti-convulsant	SBA-16[65]
Rufinamide	Anti-convulsant	SBA-16[65]

Sol-Gel Technologies' formulation platform is based upon active encapsulation within a multitude of layers of porous silica and is built around a number of the concepts and theories introduced in Section 4.3.2.2. Quaternary ammonium salts feature heavily in this aqueous, template-based preparation, as does careful control of the solution pH to facilitate condensation of the silicate units anchored to the growing formulation particles.

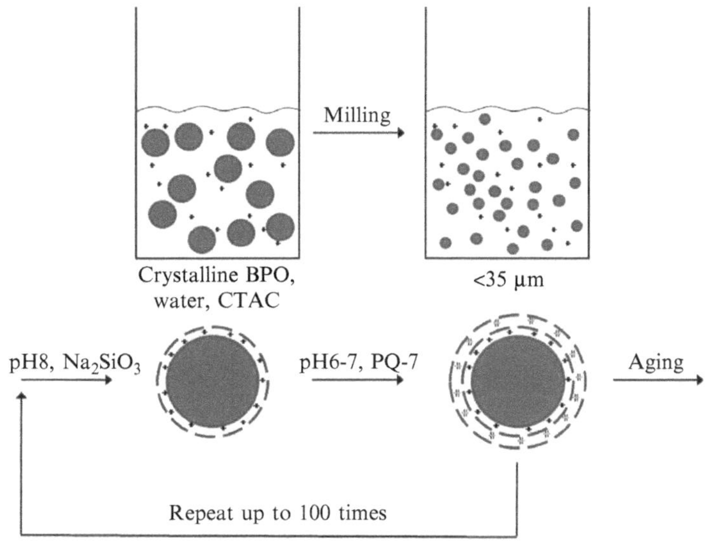

Figure 4.23 Schematic for production of benzoyl peroxide encapsulated porous microspheres.

In a typical encapsulation procedure, shown schematically in Figure 4.23, the formulations are prepared by milling a suspension of crystalline benzoyl peroxide and cetrimonium chloride in water until a $d(0.9)$ of <35 μm is achieved. The pH of this suspension is then raised to 8 using sodium hydroxide, which is followed by the addition of sodium silicate solution. The silicate anions are attracted to the cationic surfactant and at the higher pH the silicate anions polymerize to form a silicate shell. A copolymer of acrylamide and diallyldimethylammonium chloride called polyquaternium-7 is then added and the pH lowered to 6-7. The pH is then increased to pH 8 and further sodium silicate solution added to form a new layer of porous silica. This process is repeated up to 100 times to form a benzoyl peroxide core surrounded by a multitude of porous silica shells. The structure is then aged at pH 6.5 at 25 °C to consolidate the silica portions of the microshells.

One of the benefits of administering benzoyl peroxide in this way is that the benzoyl peroxide crystals are not in continuous direct contact with the skin, mitigating many of the adverse reactions that occur at the skin surface. In addition, the sustained release of benzoyl peroxide through the porous silica shells extends the effective dosage timeframe making the treatment more potent for longer periods of time.

Sol-Gel Technologies' formulation compares well against two other commercially available formulations, Benzac® AC and NeoBenzMicro.

Approximately 53% benzoyl peroxide is released after 1 hour as compared with ~95% from the other two formulations.

Despite much of the academic-based research into porous silica as a controlled release drug delivery system being focussed on oral formulations, it is relatively unsurprising that the breakthrough commercial products are topical formulations. There are far fewer development issues to overcome with topical formulations such as first pass metabolism, interactions with vascular enzymes, non-specific site of action, and transport through the necessary biological barriers/membranes to name just a few. The development of a topical formulation does not need to consider many of the potential barriers to success mentioned previously and as such proving efficacy of treatment, optimization of formulation parameters, and the route to regulatory approval are potentially all much shorter.

However, the use of porous silica for oral and parenteral treatments is such an active field of research that it is only a matter of time before a novel treatment featuring the concepts and theories introduced in this chapter successfully navigates its way through the clinical and regulatory approvals process.

REFERENCES

1. Hench LL, Splinter RJ, Greenlee TK, Allen WC. Bonding mechanisms at the interface of ceramic prosthetic materials. *J Biomed Mater Res* 1971;**5**(2):117–41.
2. Ebelman M. Recerches sur les combinaisons des acides borique et silicique avec les éthers. *Ann Chim Phys* 1846;**16**:129–66.
3. Ebelman M. *C R Acad Sci* 1847;**25**:54.
4. Iler RK. *The chemistry of silica*. New York: Wiley; 1979.
5. Stöber W, Fink A, Bohn E. Controlled growth of monodisperse silica spheres in the micron size range. *J Colloid Interf Sci* 1968;**26**(1):62–9.
6. Brinker CJ, Keefer KD, Schaefer DW, Ashley CS. Sol-gel transition in simple silicates. *J Non-Cryst Solids* 1982;**48**:47–64.
7. Yanagisawa T, Shimizu T, Kuroda K, Kato C. The preparation of alkyltrimethylammonium-kanemite complexes and their conversion to microporous materials. *Bull Chem Soc Jpn* 1990;**63**(4):988–92.
8. Beck JS, Vartuli JC, Roth WJ, Leonowicz ME, Kresge CT, Schmitt KD, et al. A new family of mesoporous molecular sieves prepared with liquid crystal templates. *J Am Chem Soc* 1992;**114**:10834–43.
9. Zhao D, Feng J, Huo Q, Melosh N, Fredrickson GH, Chmelka BF, et al. Triblock copolymer syntheses of mesoporous silica with periodic 50 to 300 Angstrom pores. *Science* 1998;**279**(5350):548–52.
10. Hansen CM. *The three dimensional solubility parameter and solvent diffusion coefficient and their importance in surface coating formulation*. Copenhagen: Danish Technical Press; 1967.
11. Hildebrand JH, Scott RL. *The solubility of nonelectrolytes*. 3rd ed. New York, NY: Dover Publications; 1964.
12. Brandrup J, Immergut EH. *Polymer handbook*. 3rd ed. New York: Wiley Interscience; 1989.

13. Flory PJ. Thermodynamics of high polymer solutions. *J Chem Phys* 1941;**9**(8):660–1.
14. Flory PJ. Thermodynamics of high polymer solutions. *J Chem Phys* 1942;**10**(1):51–60.
15. Huggins ML. Solutions of long chain compounds. *J Chem Phys* 1941;**9**:440.
16. Huggins ML. Some properties of solutions of long-chain compounds. *J Phys Chem* 1942;**46**:151–8.
17. Templin M, Franck A, Du Chesne A, Leist H, Zhang Y, Ulrich R, et al. Organically modified aluminosilicate mesostructures from block copolymer phases. *Science* 1997;**278**(5344):1795–8.
18. Brinker CJ, Lu Y, Sellinger A, Fan H. Evaporation-induced self-assembly: nanostructures made easy. *Adv Mater* 1999;**11**(7):579–85.
19. Kresge CT, Leonowicz ME, Roth WJ, Vartuti JC, Beck JS. Ordered mesoporous molecular sieves synthesized by a liquid-crystal template mechanism. *Nature* 1992;**359**(6397):710–2.
20. Huo Q, Margolese DI, Ciesla U, Feng P, Gier TE, Sieger P, et al. Generalised synthesis of periodic surfactant/inorganic composite materials. *Nature* 1994;**368**(6469):317–21.
21. Bagshaw A, Prouzet E, Pinnavia TJ. Templating of mesoporous molecular sieves by nonionic polyethylene oxide surfactants. *Science* 1995;**269**(5229):1242–4.
22. Huo Q, Margolese DI, Stucky GD. Surfactant control of phases in the synthesis of mesoporous silica-based materials. *Chem Mater* 1996;**8**(5):1147–60.
23. Zhao D, Huo Q, Feng J, Chmelka BF, Stucky GD. Nonionic triblock and star diblock copolymer and oligomeric surfactant syntheses of highly ordered, hydrothermally stable, mesoporous silica structures. *J Am Chem Soc* 1998;**120**(24):6024–36.
24. Zhao D, Yang P, Chmelka BF, Stucky GD. Multiphase assembly of mesoporous-macroporous membranes. *Chem Mater* 1999;**11**(5):1174–8.
25. Smarsly B, Polarz S, Antonietti M. Preparation of porous silica materials via sol-gel nanocasting of nonionic surfactants: a mechanistic study on the self-aggregation of amphiphiles for the precise prediction of the mesopore size. *J Phys Chem B* 2001;**105**(43):10473–83.
26. Tadjarodi A, Haghverdi M, Mohammadi V. Preparation and characterization of nanoporous silica aerogel from rice husk ash by drying at atmospheric pressure. *Mater Res Bull* 2012;**47**:2584–9.
27. Horcajada P, Rámila A, Férey G, Vallet-Regí M. Influence of superficial organic modification of MCM-41 matrices on drug delivery rate. *Solid State Sci* 2006;**8**:1243–9.
28. de Sousa A, de Sousa EMB. Ordered mesoporous carrier silica system applied in nanobiotechnology. *Braz Arch Biol Technol* 2005;**48**:243–50.
29. Levene L, Thomas IL. U.S. Patent 3,640,093; February 8, 1972.
30. Dislich H. New routes to multicomponent oxide glasses. *Angew Chem Int Ed* 1971;**10**(6):363–70.
31. Schilde C, Hanisch C, Naumann D, Beierle T, Kwade A. A novel way to vary the structure of precipitated silica and calcium carbonate aggregates in a wide range by using grinding media during the precipitation process. *Chem Eng Sci* 2013;**94**:127–37.
32. Schlomach J, Kind M. Investigations on the semi-batch precipitation of silica. *J Colloid Interf Sci* 2004;**277**:316–26.
33. Aw MS, Simovic S, Yu Y, Addai-Mensah J, Losic D. Porous silica microshells from diatoms as biocarrier for drug delivery applications. *Powder Technol* 2012;**223**:52–8.
34. Mikulasik et al. Method for purifying diatomaceous earth suitable for pharmaceutical use. European patent EP2627606 A2; 2013.
35. Monkhouse DC, Lach JL. Use of adsorbents in enhancement of drug dissolution I. *J Pharm Sci* 1972;**61**(9):1430–5.
36. Yang KY, Glemza R, Jarowski CI. Effects of amorphous silicon dioxides on drug dissolution. *J Pharm Sci* 1979;**68**(5):560–5.
37. Narurkar AN, Jarowski CI. Dissolution rates of prednisolone triturations using porous and non-porous silicas. I. Pilot oral absorption study in dogs. *Drug Dev Ind Pharm* 1983;**9**(6):999–1010.

38. Braun S, Rappoport S, Zusman R, Avnir D, Ottolenghi M. Biochemically active sol-gel glasses: the trapping of enzymes. *Mater Lett* 1990;**10**(1-2):1–5.
39. Ellerby LM, Nishida CR, Nishida F, Yamanaka SA, Dunn B, Valentine JS, et al. Encapsulation of proteins in transparent porous silicate glasses prepared by the sol-gel method. *Science* 1992;**255**(5048):1113–5.
40. Nicoll SB, Radin S, Santos EM, Tuan RS, Ducheyne P. *In vitro* release kinetics of biologically active transforming growth factor-β1 from a novel porous glass carrier. *Biomaterials* 1997;**18**:853–9.
41. Lyons RM, Keski-Oja J, Moses HL. Proteolytic activation of latent transforming growth factor-β from fibroblast-conditioning medium. *J Cell Biol* 1988;**106**:1659–65.
42. Sieminska L, Zerda TW. Diffusion of steroids from sol-gel glass. *J Phys Chem* 1996;**100** (11):4591–7.
43. Koskinen. *Proceedings of the 30th annual meeting on controlled release society*. UK: Glasgow; 2003.
44. Mahoney D, Cavallaro AS, Mody KT, Xiong L, Mahoney TJ, Qiao SZ, et al. *In vivo* delivery of bovine viral diarrhoea virus, E2 protein using hollow mesoporous silica nanoparticles. *Nanoscale* 2014;**6**:6617–26.
45. Pope EJA, Braun K, Peterson CM. Bioartificial organs I: silica gel encapsulated pancreatic islets for the treatment of diabetes mellitus. *J Sol-Gel Sci Technol* 1997;**8**(1-3):635–9.
46. Al-Saraj M, Abdel-Latif MS, El-Nahal I, Baraka R. Bioaccumulation of some hazardous metals by sol-gel entrapped microorganisms. *J Non-Cryst Solids* 1999;**248** (2-3):137–40.
47. Sglavo VM, Carturan G, Dal Monte R, Muraca M. SiO_2 entrapment of animal cells. Part I: mechanical features of SiO_2 coatings. *J Mater Sci* 1999;**34**(15):3587–90.
48. Conroy JFT, Power ME, Martin J, Earp B, Hosticka B, Daitch CE, et al. Cells in sol-gels. I: a cytocompatible route for the production of macroporous silica gels. *J Sol-Gel Sci Technol* 2000;**18**(3):269–83.
49. Gu J, Huang K, Zhu X, Li Y, Wei J, Zhao W, et al. Sub-150 nm mesoporous silica nanoparticles with tunable pore sizes and well-ordered mesostructure for protein encapsulation. *J Colloid Interf Sci* 2013;**407**:236–42.
50. dos Santos SML, Nogueira KAB, Gama MS, Lima JDF, Júnior IJS, de Azevedo DCS. Synthesis and characterization of ordered mesoporous silica (SBA-15 and SBA-16) for adsorption of biomolecules. *Microporous Mesoporous Mater* 2013;**180**:284–92.
51. Avnir D, Braun S, Lev O, Ottolenghi M. Enzymes and other proteins entrapped in sol-gel materials. *Chem Mater* 1994;**6**(10):1605–14.
52. Santos EM, Radin S, Ducheyne P. Sol-gel derived carrier for the controlled release of proteins. *Biomaterials* 1999;**20**(18):1695–700.
53. Bhatia RB, Brinker CJ. Aqueous sol-gel process for protein encapsulation. *Chem Mater* 2000;**12**(8):2434–41.
54. Keeling-Tucker T, Rakic M, Spong C, Brennan JD. Controlling the material properties and biological activity of lipase within sol-gel derived bioglasses via organosilane and polymer doping. *Chem Mater* 2000;**12**(12):3695–704.
55. Flora KK, Brennan JD. Effect of matrix aging on the behaviour of human serum albumin entrapped in a tetraethyl orthosilicate-derived glass. *Chem Mater* 2001;**13**(11):4170–9.
56. Viitala R, Jokinen M, Rosenholm JB. Mechanistic studies on release of large and small molecules from biodegradable SiO_2. *Int J Pharm* 2007;**336**:382–90.
57. Popova M, Szegedi A, Mavrodinova V, Tušar N, Mihály J, Klébert S, et al. Preparation of resveratrol-loaded nanoporous silica materials with different structures. *J Solid State Chem* 2014;**219**:37–42.
58. Tušar N, Ristić A, Mali G, Mazaj M, Arčon I, Arčon D, et al. MnOx nanoparticles supported on a new mesostructured silicate with textural porosity. *Chem Eur J* 2010;**16** (19):5783–93.

59. Wu C, Sun W, Zhao Z, Zhao Y, Hao Y, Liu Y, et al. Synthesis of novel core-shell structured dual-mesoporous silica nanospheres and their application for enhancing the dissolution rate of poorly water soluble drugs. *Mater Sci Eng C* 2014;http://dx.doi.org/10.1016/j.msec.2014.08.040.
60. Kumar D, Sailaja Chirravuri SV, Shastri NR. Impact of surface area of silica particles on dissolution rate and oral bioavailability of poorly water soluble drugs: a case study with aceclofenac. *Int J Pharm* 2014;**461**:459–68.
61. Zhu W, Wan L, Zhang C, Gao Y, Zheng X, Jiang T, et al. Exploitation of 3D face-centered cubic mesoporous silica as a carrier for a poorly water soluble drug: influence of pore size on release rate. *Mater Sci Eng C* 2014;**34**:78–85.
62. Uejo F, Limwikrant W, Moribe K, Yamamoto K. Dissolution improvement of fenofibrate by melting inclusion in mesoporous silica. *Asian J Pharm* 2013;**8**:329–35.
63. Van Speybroeck M, Mellaerts R, Mols R, Thi TD, Martens JA, Van Humbeeck J, et al. Enhanced absorption of the poorly soluble drug fenofibrate by tuning its release rate from ordered mesoporous silica. *Eur J Pharm Sci* 2010;**41**:623–30.
64. Chen B, Quan G, Wang Z, Chen J, Wu L, Xu Y, et al. Hollow mesoporous silicas as a drug solution delivery system for insoluble drugs. *Powder Technol* 2013;**240**:48–53.
65. Thomas MJK, Slipper I, Walunj A, Jain A, Favretto ME, Kallinteri P, et al. Inclusion of poorly soluble drugs in highly ordered mesoporous silica nanoparticles. *Int J Pharm* 2010;**387**:272–7.
66. Cao X, Deng W, Fu M, Zhu Y, Liu H, Wang L, et al. Seventy-two-hour release formulation of the poorly soluble drug silybin based on porous silica nanoparticles: In vitro release kinetics and in vitro/in vivo correlations in beagle dogs. *Eur J Pharm Sci* 2013;**48**:64–71.
67. Guo Z, Liu XM, Ma L, Li J, Zhang H, Gao YP, et al. Effects of particle morphology, pore size and surface coating of mesoporous silica on Naproxen dissolution rate enhancement. *Colloid Surf B* 2013;**101**:228–35.
68. Jia L, Shen J, Li Z, Zhang D, Zhang Q, Duan C, et al. Successfully tailoring the pore size of mesoporous silica nanoparticles: exploitation of delivery systems for poorly water-soluble drugs. *Int J Pharm* 2012;**439**:81–91.
69. Cao X, Fu M, Wang L, Liu H, Deng W, Qu R, et al. Oral bioavailability of silymarin formulated as a novel 3-day delivery system based on porous silica nanoparticles. *Acta Biomater* 2012;**8**:2104–12.
70. Hu Y, Wang J, Zhi Z, Jiang T, Wang S. Facile synthesis of 3D cubic mesoporous silica microspheres with a controllable pore size and their application for improved delivery of a water-insoluble drug. *J Colloid Interf Sci* 2011;**363**:410–7.
71. Hu Y, Zhi Z, Zhao Q, Wu C, Zhao P, Jiang H, et al. 3D cubic mesoporous silica microsphere as a carrier for poorly soluble drug carvedilol. *Microporous Mesoporous Mater* 2012;**147**:94–101.
72. Kinnari P, Mäkilä E, Heikkilä T, Salonen J, Hirvonen J, Santos HA. Comparison of mesoporous silicon and non-ordered mesoporous silica materials as drug carriers for itraconazole. *Int J Pharm* 2011;**414**:148–56.
73. Mellaerts R, Mols R, Jammaer JAG, Aerts CA, Annaert P, Van Humbeeck J, et al. Increasing the oral bioavailability of the poorly water soluble drug itraconazole with ordered mesoporous silica. *Eur J Pharm Biopharm* 2008;**68**:223–30.
74. Tingming F, Liwei G, Kang L, Tianyao W, Jin L. Template occluded SBA-15: an effective dissolution enhancer for poorly water-soluble drug. *Appl Surf Sci* 2010;**256**:6963–8.
75. Zhang Y, Zhi Z, Jiang T, Zhang J, Wang Z, Wang S. Spherical mesoporous silica nanoparticles for loading and release of the poorly water-soluble drug telmisartan. *J Control Release* 2010;**145**:257–63.
76. Sertchook H, Toledano O, Bar-Simantov H. Sol-Gel Technologies, compositions for the treatment of Rosacea. WO 2014/083557 A1; 27th November 2012.

CHAPTER 5

Mesoporous Bioactive Glass-Based Controlled Release Systems

Contents

5.1 Introduction 139
5.2 General Synthesis Methods for Mesoporous Bioactive Glasses 140
5.3 Mesoporous Glass-Based 3D Scaffold Structures 141
5.4 The Role of MBG Surface Functionalization for Controlling Active Release 146
5.5 Case Studies 148
5.5.1 Delivery of therapeutic ions from MBGs 149
5.5.2 Delivery of drugs and growth factors from MBGs 151
5.5.3 Delivery of drugs from a polymer coating of MBG 153
5.6 Conclusions and Future Context 154
References 154

5.1 INTRODUCTION

Due to their superior biocompatibility, osteostimulating properties, high degradation rate and angiogenic potential, bioactive glasses play a central role as bone substituting materials.[1–7] In 2004, a new family of sol-gel glasses with tailored porosity at the nanometre scale was developed by Yan et al.[8–10] who combined the supramolecular chemistry of surfactants and the sol-gel method.[1,10,11] These materials have the same or similar compositions of glasses produced by more traditional fusion methods, but with the added ability to introduce designed mesoporosity. As a result, new directions and possibilities are now available for applying nanotechniques to regenerative medicine by coupling drug delivery with a bioactive inorganic material.[1,12,13] Mesoporous bioactive glasses (MBGs), based on a $CaO-SiO_2-P_2O_5$ composition [refer to Chapter 2 on the materials fundamentals of inorganic controlled release technology (inorganic CRT) systems], are characterized by a highly ordered mesopore channel structure

Inorganic Controlled Release Technology
http://dx.doi.org/10.1016/B978-0-08-099991-3.00005-3

with monodisperse pore sizes ranging from 2 to 50 nm, and have attracted increasing attention in the last 10 years due to their high potential for applications in drug delivery and tissue engineering.[1,14–20] The large surface area of MBGs, their controlled mesostructure, well-defined surface properties, facile modification and tuneable pore size and volume enhance the bioactive behaviour and allow them to be loaded with biologic or pharmaceutical agents to promote new bone formation.[8,12,19,21–26] In addition, they can be used in biomedical applications, for example, as scaffolds for bone tissue engineering (BTE)[8,9,16,21,24,27–29] and drug delivery systems (DDS).[9,19,21,30–33] More recently, DDS based on mesoporous materials have become an extensively studied class of materials.[19–21,34] Vallet-Regí and coworkers have focussed specifically on mesoporous silica as a particularly attractive material considering its ordered mesoporous structure, good biocompatibility, low cytotoxicity, tailored surface charge and wide range of organic functionalization.[9,16,19,21,22,35–38] Indeed, MBGs can be considered as a new class of inorganic CRT functional biomaterials which combine bioactivity and efficient drug delivery properties.[21,39] Some of the fundamental features of MBGs and an overview of their recent development as controlled drug release systems will be highlighted and discussed.

5.2 GENERAL SYNTHESIS METHODS FOR MESOPOROUS BIOACTIVE GLASSES

The preparation of MBGs is similar to that for mesoporous SiO_2 in which supramolecular chemistry has been incorporated into the sol-gel process.[1,8–11] When designing controlled release systems based on MBGs, it is important to consider the physicochemical properties of the functional molecule to be encapsulated and subsequently delivered, and how the characteristics of the MBG carrier system can be modified to suit as appropriate. Typically, for the delivery of most common small-molecule compounds, non-ionic block copolymers such as Pluronic® P123 and F127 are used as structure-directing agents, which are essential for obtaining well-ordered structures. The formation mechanism of the hierarchical MBGs can be generalized as follows. The initial homogeneous sol is obtained by dissolving soluble bioactive glass precursors [e.g. tetraethyl orthosilicate (TEOS), triethyl phosphate (TEP)], a surfactant and a strong acid (mostly hydrochloric acid or nitric acid) in an ethanol-water mixed solvent system. The surfactants present in the sol self-organize into micelles, which are able to link the hydrolysed precursors (e.g. TEOS, TEP) through the

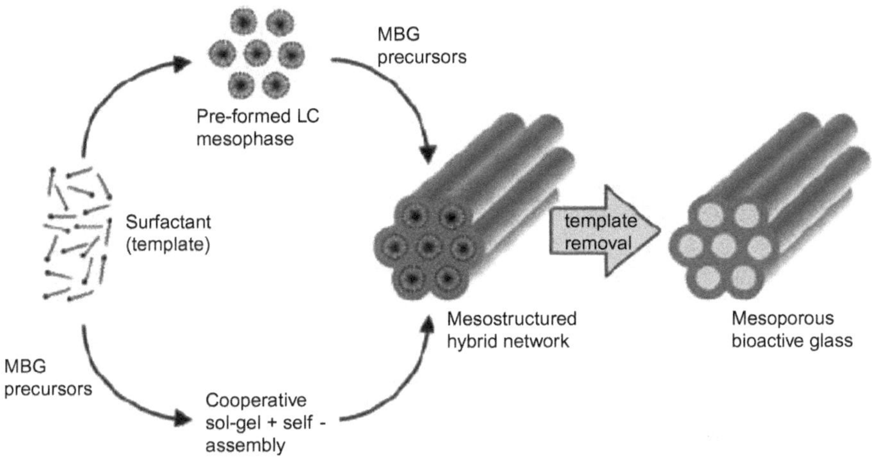

Figure 5.1 Flowchart showing mesostructure formation of a MBG. *(Reprinted from Ref 40. Copyright 2002 American Chemical Society).*

hydrophilic component and self-assemble to form an ordered mesophase, as shown in Figure 5.1.

This ordered mesophase reaction system containing the sol-gel precursors and surfactants then undergoes an evaporation-induced self-assembly (EISA) process. During this time the surfactant molecules assemble into micelles that maintain the hydrophilic parts in contact with the hydrolysed silica precursor while shielding the hydrophobic parts within the micellar core. Once the mixture is completely dried and the surfactant removed by calcination, a well-ordered mesoporous structure will be obtained, exhibiting high surface area and porosity.[1,9,11–13,21,33,35] Since temperatures are typically in the range 600-700 °C, most disadvantages associated with the high-temperature thermal processing of glasses and glass-ceramics can be eliminated.[41]

Moreover, sol-gel based methods also offer potential advantages for powder production due to enhanced and better control of bioactivity by changing either the composition or the microstructure through processing parameters.[9,42] A schematic representation of the synthesis of MBGs through the EISA method is shown in Figure 5.2.

5.3 MESOPOROUS GLASS-BASED 3D SCAFFOLD STRUCTURES

MBGs can also be prepared as three-dimensional (3D) porous scaffolds. Sepulveda et al.[43] described the preparation of macroporous scaffolds by

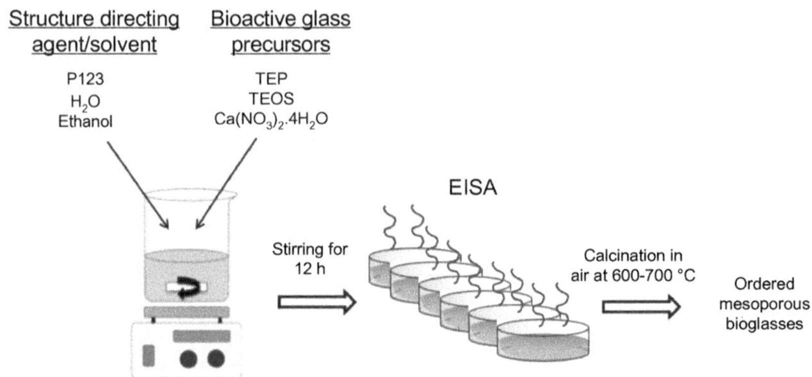

Figure 5.2 Schematic representation of synthesis of MBGs through the EISA method.

foaming of a sol-gel system. A scaffold's structure must consist of a fully interconnected porous network with pores >100 μm connected by channels of smaller size to enable complete access of nutrients to the cells to support vascularization, allow tissue ingrowth and induce favourable integration with host tissues once implanted.[43,44] Processing of sol-gel systems in this way allows the fabrication of these characteristic foam structures which have a porosity hierarchy resembling that of trabecular bone.[43,44] The different synthesis steps are illustrated in Figure 5.3.

The direct foaming procedure approach produces porous bioactive glass scaffolds with interconnectivity suitable for vascularized bone ingrowth with

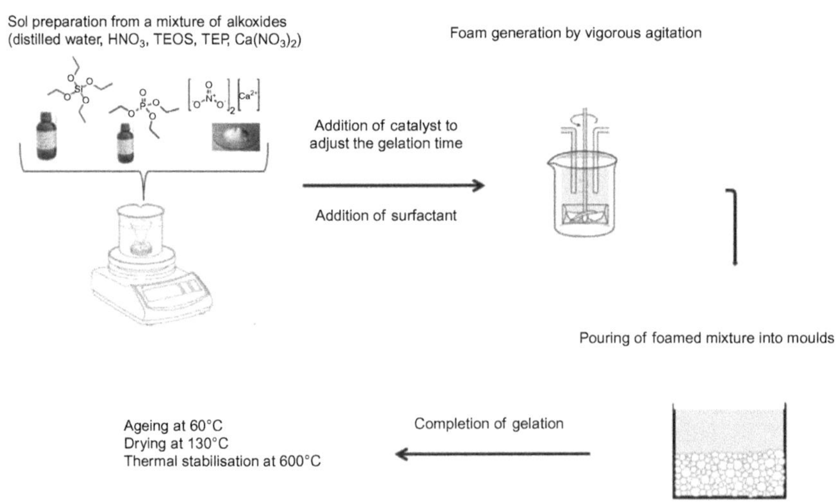

Figure 5.3 Schematic representation of the manufacture of sol-gel foams.[43]

compressive strengths similar to those of cancellous bone. Typical features of sol-gel-derived scaffolds are the mesoporous structure and high surface area of the bioactive foam framework.[45,46] In the aforementioned study, the resulting tissue engineering scaffolds contain 3D hierarchical structures of an interconnected macroporous network in a matrix also consisting of mesopores with pore diameters ranging from 10 to 30 nm. The benefits of this enhanced nanoporosity include providing controlled degradation behaviour as well as sites for cell attachment and protein adsorption.[47]

As an extension of this scaffold-forming process Yun et al.[33] used methylcellulose (MC) as a porogen to obtain porous MBG scaffolds with large pores of 100 μm. A sol solution containing silicon, calcium and phosphorus inorganic precursors and the non-ionic block copolymer structure-directing agent P123 (or F127) was first prepared to produce the self-assembling hexagonal- (or cubic-) mesostructures in solution. In order to introduce macropores, MC was then added to this precursor solution, after which the ordered mesostructure was found to be still well maintained.[33] Following typical thermal treatment, a product mesostructured bioglass was produced which contained additional macroporosity (Figure 5.4).

It was highlighted by the authors that mesostructured bioactive glass particles are effective at dispersing the MC whilst in solution, resulting in a homogeneously distributed and well-interconnected macrostructure. In this way, MBG products with relative porosities of up to 75-90% can be obtained. The synthesized scaffolds exhibited good compressive strength, superior *in vitro* bioactivity and favourable biocompatibility.[33]

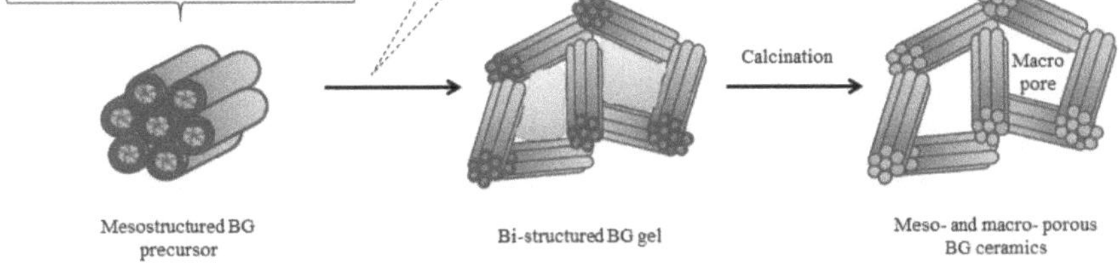

Figure 5.4 Synthesis of hierarchically porous bioactive glasses according to the procedure introduced by Yun et al.[33] *(Reproduced from Ref 33 with permission of John Wiley & Sons).*

The replication technique, which is widely used as a method for the fabrication of 3D porous structures, using polyurethane (PU) sponges as sacrificial templates, was applied for the first time in combination with MBG in 2007 by Wu et al.[48] (four different MBG compositions: MBGs 70S25C, MBGs 80S15C, MBGs 90S5C and MBGs 100S) and Li et al.[49] (composition MBGs 80S15C). Both groups prepared MBG following the method reported by Yan et al.[8] Afterwards, the PU sponges were completely immersed in the sol, the excess of which was then squeezed out and the foam left at room temperature for the remaining solution to evaporate. When using a 3D template in this way the sol permeates inside the struts of the reticulated foam because of its porous structure. After repeating this procedure six times, the coated sponges were calcined at 700 °C, applying a slow heating rate and kept at this temperature for 5 h. After the macro- and mesotemplates are burned out, undergoing a slow heating rate, a positive replica of the macro-template with mesoporous structured bioactive glass struts is obtained.[49,50] The so fabricated MBG scaffolds exhibited an interconnected macroporous network with pore sizes of 200-400 μm and uniform mesoporous walls with pores ranging in size from 3.6 to 4.9 nm. Materials prepared by this route are effective at stimulating HCA deposition on their surfaces after only 4 h in simulated body fluid.[48,49] Further developments of varied compositions of MBG scaffolds for BTE and drug delivery applications were made and studied by Wu's group.[51–54]

In similar studies, Wang et al.[55] fabricated MBG scaffolds with hierarchical pore structures using a natural Mediterranean Sea sponge as the macropore template and P123 as the mesopore template. Following calcinations at 580 °C for 6 h, the obtained MBG scaffolds were highly porous with pore sizes ranging from 2 to 5 μm, and also featured sustained DDS properties.[55] Another study was published by Lin et al.,[50] who successfully synthesized hierarchical mesoporous-macroporous bioactive glass scaffolds by a cotemplated approach, where the stem core of corn was used as the macroporous biotemplate and P123 as the mesoporous template. The material exhibited interconnected macropores, with 60-120 μm pore sizes, uniform mesopores of 4.6 nm as well as excellent bioactivity.

The advantages of the replication technique are the ability to obtain highly interconnected porous structures whilst also having a high level of control over the pore size and porosity. Nevertheless, these materials' main disadvantage is the low mechanical strength,[1] which is a critical obstacle for the application as BTE scaffolds in load-bearing applications.

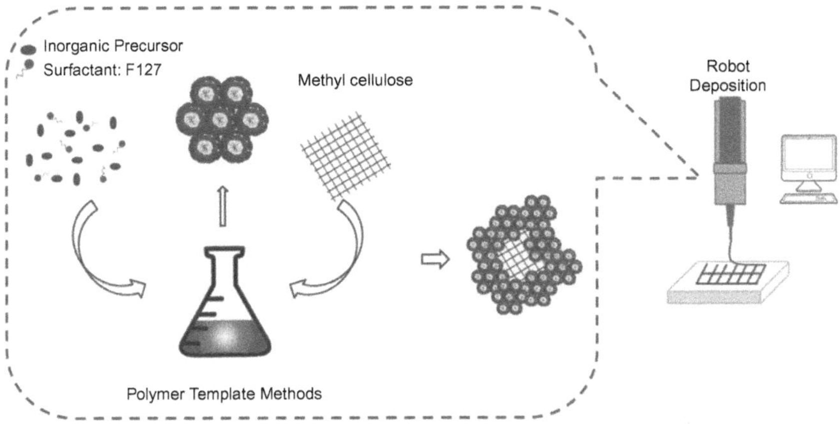

Figure 5.5 Schematic drawing of synthetic concept of hierarchical 3D pore structure MBG scaffold, according to the processing method introduced by Yun et al.[56] *(Reprinted with permission from Ref 55. Copyright 2007 American Chemical Society).*

MBG scaffolds can also be produced by a 3D plotting technique, also called direct printing or robocasting. This method allows a better control of pore morphology, pore size and porosity. The studies of Yun et al.[56] and Garcia et al.[57] on hierarchical porous MBG scaffolds illustrate the direct printing technique. MBG gel, where the mesopores were made using F127, was mixed with MC template, inducing macrosized pores with a size range between 10 and 100 μm.[56,57] This mixture was then printed, creating giant-sized pores (>100 μm) controlled by the computer system, as illustrated in Figure 5.5. In a next step, scaffolds were sintered at 500-700 °C to remove polymer templates and obtain the final calcined porous scaffolds. Yun et al.[56] also presented MBG scaffolds with outstanding bone-forming activity *in vitro*.

This technique allows the architectures of the scaffolds to be accurately controlled by a layer-by-layer plotting under mild conditions,[1,58–60] which represents a significant advantage over other methods. However, a limitation is the incorporation of the MC, resulting in some micropores, which compromise the scaffolds' mechanical strength.[1]

In 2011, Wu et al.[61] reported an alternative approach using a modified 3D printing technique, introducing polyvinylalcohol as binder. They successfully achieved 3D printed MBG scaffolds having reduced porosity (60.4%), improved mechanical strength (approximately 200 times that of scaffolds prepared by the replication technique which normally have a porosity of ~90%), controllable pore architecture, excellent bioactivity and sustained drug delivery properties.[61] As these structures are fabricated from bioactive glass, they demonstrate the positive attributes of resorbability, bioactivity and biocompatibility. These scaffolds also fulfil the criterion of being osteogenic as they stimulate human primary osteoblasts to produce mineralized bone matrix *in vivo* without adding growth factors or hormones to the cell culture, needed in the case of other materials.[5,43,47] As an extra advantage these mesoporous systems can be loaded with different drugs or bioactive molecules to develop flexible and controllable release systems.

5.4 THE ROLE OF MBG SURFACE FUNCTIONALIZATION FOR CONTROLLING ACTIVE RELEASE

The possibility of developing inorganic CRT systems with ordered mesoporous structure represents an important advance for the biomaterials community. MBGs exhibit similar compositions to conventional sol-gel-derived glasses (CaO-SiO_2-P_2O_5) with highly ordered mesoporous arrangements and because of their high surface area they also exhibit accelerated bioactive response.[1,8,9,11–13,21,35,43,48] In the framework of BTE, mesoporous materials have been proposed to be used as carriers for local delivery of bioactive molecules and therapeutic drugs.[21] The highly ordered mesoporous structure allows MBGs to be loaded with either osteogenic agents to promote new bone formation[8,16,19–21,25,26,34,49] or drugs[13,21,37,62] acting as a controlled delivery system. The advantages of this approach include high delivery efficiency, continuous release, reduced toxicity and improved patient compliance by avoidance of secondary effects resulting from systemic administration of drug. Due to the local application, disadvantages of systemic delivery can be avoided which are the inactivation by enzymes or chemical reactions in the blood.[9,11,21,63] Moreover, the presence of abundant silanol groups (refer to Chapter 2) on the mesostructure surface offers an anchoring point for organic functional groups *via* molecular interactions, such as van der Waals forces and/or strong hydrogen bonds.[9,13,21,35] All these properties are essential in the loading and release of molecules[37] and are illustrated in Figure 5.6. Additionally, silanol groups also give the possibility to functionalize the mesoporous walls with different organic groups *via* chemical methods.[64] To proceed with this functionalization, two main routes are used. One is the post-synthesis surface functionalization including silylation or grafting, whilst the introduction of extra functionality by co-condensation of alternative sol-gel precursors is the other.[9,13]

The post-synthesis functionalization process is performed by grafting the organic functions to the previously formed pure inorganic mesoporous silica

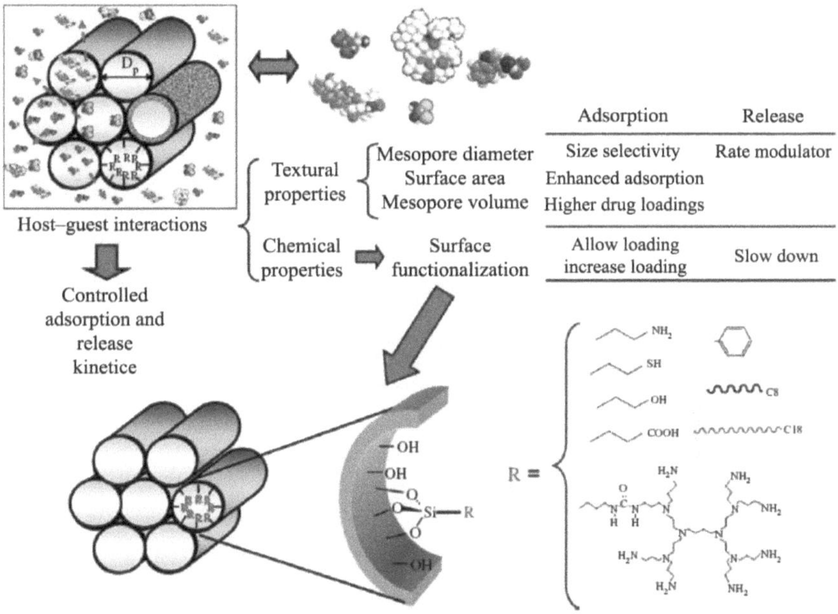

Figure 5.6 Parameters governing loading and release rate of drug molecules in silica-based mesoporous materials. *(Reproduced from Ref. 13 with permission of The Royal Society).*

matrix, that is, by reaction of organosilanes $(R'O)_3SiR$ with the free silanol groups of the pore surface, under anhydrous conditions.[9,13] In this way it is possible to retain the mesostructure of the starting material, although there might be a reduction in the porosity due to the presence of the organic moiety. The main disadvantage of this pathway is that some organosilanes preferentially react at the pore entrance, which may reduce the entrance of further organosilanes inside the pores. This could lead to a non-homogeneous distribution of the organic groups inside the porous structure.[9] The co-condensation method involves the simultaneous condensation of the corresponding silica and organosilica precursors, $(R'O)_3SiR$, in the presence of the structure-directing agents during the mesoporous synthesis—the functionalization process can be carried out in one step.[9,13] Particular attention should be paid during the surfactant removal to avoid destroying or degrading the organic functionalization. For this reason solvent extraction methods should be used instead of calcination. The main advantage of the co-condensation method is the more homogeneous distribution of the organic functionalities inside the pore walls because the organic units are direct components of the MBG matrix. At the same time an increase of the organosilane amount in the reaction mixture could affect or totally destroy the mesoporous order. Also, by increasing the amount of organic functionalities present, an important reduction of pore diameter, pore volume and specific surface area has been observed.[9] The maximum achievable degree of organic modification varies depending upon the choice of functionalization route. The co-condensation method, leading to disordered products, does not achieve more than 40 mol% of organic functionality content in the modified silica phases.[64] Contrarily, the post-synthesis method enables a higher functionalization rate but the introduced functional groups are not homogeneously distributed.[60]

The surface functionalized mesoporous silicate materials containing the organic anchoring points will interact with the guest molecules *via* the attracting intermolecular bonding forces including van der Waals and the various electrostatic-based interactions. These specific host-guest molecule interactions can be varied depending upon the carrier surface functionalization and specific microstructure of the DDS, which in turn allows fine tuning of the drug loading and release kinetics.

5.5 CASE STUDIES

The treatment of bone defects, especially large bone defects due to trauma, infections, tumours, or genetic malformations, represents a major challenge for clinicians.[65] In fact, large bone defects lack the template for an orchestrated regeneration and require surgical intervention. For better regeneration of bone, bioactive scaffolds should not only be osteoconductive (to support bone growth and encourage the ingrowth of surrounding bone) but they should also induce and support osteogenesis (formation of new bone) and angiogenesis (for inducing neovascularization).[9,11,66] Moreover, inflammatory responses or infections, which are associated with orthopaedic implants or bone replacements, are ever-present problems which can lead to serious clinical complications.[11,67] Conventional treatments include administration of antibiotics, surgical debridement and implant removal. However, these approaches have limitations and could result in additional surgical interventions.[68] As mentioned above, a convenient approach to overcome this problem is the introduction of a local DDS into the implant site.[13,21] In order to address these significant clinical issues, it is important to design multifunctional bioactive scaffolds by introducing ions, growth factors and therapeutic drugs.[11] The interest in MBGs has increased over the last few years due to their ordered mesostructure and composition-related beneficial properties, which makes these materials excellent candidates for inorganic DDS.[1,9,11]

5.5.1 Delivery of Therapeutic Ions from MBGs

The incorporation of therapeutic ions in bioactive glasses for enhanced bone regeneration is an expanding field.[69] In the past 5 years different therapeutic ions have predominantly been incorporated into MBGs by an *in situ* synthesis method. In this process, the structure-directing agent is firstly dissolved in an ethanolic solution followed by the therapeutic ion precursors (Cu^{2+}-Cu $(NO_3)_2$; Ag^+-$AgNO_3$; Li^+-$LiNO_3$ amongst others) and the glass network-forming precursors.[51,52,70–79] The reaction mixture is then allowed to undergo an EISA process which is followed by a calcination step. The resulting material is a therapeutic ion-doped MBG. In this way the ions are part of the glass framework and the ordered porous structure is maintained. These ions, once released, can have significant osteogenic, cementogenic, angiogenic and antibacterial effects, as summarized in Table 5.1.

Wu et al.[80] reported the development of hypoxia-mimicking MBG scaffolds by incorporating ionic Co^{2+} and/or ionic Cu^{2+} into MBG scaffolds and by this incorporation, enhancing the hypoxic function. Hypoxia (low oxygen pressure) plays a main role in coupling angiogenesis with osteogenesis.[70,80] Hypoxia activates a series of angiogenic processes mediated by hypoxia inducing factor-1a (HIF-1α) transcription factor. HIF-1α initiates

Table 5.1 The positive effect of therapeutic ion release from MBG on osteogenesis, cementogenesis, angiogenesis and antibacterial activities[51,52,70–79]

Therapeutic ions released from MBG	Functional properties				Inhibition of osteoclast activity
	Osteogenesis	Cementogenesis	Angiogenesis	Antibacterial activity	
Ag^+	✓			✓	
Co^{2+}	✓		✓		
Cu^{2+}	✓		✓	✓	
Li^+	✓	✓			
Sr^{2+}	✓	✓			
Zn^{2+}	✓			✓	
Mg^{2+}	✓				
B^{3+}	✓				
Ga^{3+}					✓
Ce^{3+}	✓				

the expression of a number of genes associated with tissue regeneration, skeletal tissue development and enhances fracture repair.[70] Co^{2+} and Cu^{2+} are known to chemically induce HIF-1α to promote a hypoxia-like response. Moreover, Wu et al. demonstrated that Cu-MBG significantly inhibited bacterial viability (e.g. *Escherichia coli*) compared to blank control scaffolds.[70,75,77] In addition, Ag-containing MBG scaffolds showed a very good antibacterial effect against *Staphylococcus aureus* (*S. aureus*), with strong evidence of bactericidal activity with a glass concentration as low as 0.5 mg/ml.[71] Ionic Li^{+} has been widely used as a long-term mood stabilizer in the treatment of bipolar and depressive disorders.[78] In 2005, Clément-Lacroix et al.[81] reported that Li^{+} is also able to increase bone formation and bone mass *in vivo*. Han et al.[78] prepared for the first time Li-MBG scaffolds, with large-sized pores ranging from 300 to 500 μm. Five percent (mol%) of Li^{+} incorporated in the MBG scaffolds enhanced the proliferation and osteogenic/cementogenic differentiation of human periodontal ligament cells (hPDLCs), which has great clinical relevance, especially for periodontal tissue engineering (e.g. regeneration of alveolar bone, root cementum and periodontal ligament). Divalent ions, such as Sr^{2+}, Zn^{2+} and Mg^{2+}, once released from MBGs can stimulate the osteogenic/cementogenic differentiation of human bone marrow stromal cells (hBMSCs) and hPDLCs.[51,72–74,79] Sánchez-Salcedo et al.[82] showed that MGB scaffolds enriched with 4% (mol%) ZnO were favourable for human osteosarcoma cell development and that scaffolds exhibited antibacterial properties against *S. aureus*. Zhu et al.[51] prepared for the first time Zr-containing MBG scaffolds with different Zr content. The study showed that Zr incorporation not only enhanced the compressive strength of the MBG scaffolds but also decreased the rate of glass dissolution and maintained a more stable pH environment.[73] Furthermore, it provided a better environment for hBMSC attachment, spreading and proliferation on these scaffolds.[51,73] Zhang et al.[72] showed that the incorporation of Sr^{2+} into MBG scaffolds significantly stimulated new bone formation in osteoporotic bone fractures when compared to MBG scaffolds alone. Strontium in the form of strontium ranelate is already used in clinical practice as an antiresorptive agent for the treatment of osteoporosis. The advantage of strontium ranelate over other therapies is that it simultaneously stimulates bone formation and decreases bone resorption. However it still remains difficult to regenerate osteoporosis-related injuries post fractures.[72] For this reason, Sr-MBGs could be a viable method to regenerate large-sized fractures due to osteoporosis.[74]

Inclusion of trivalent ions into MBGs has also been investigated. Cerium (Ce^{3+}) has been found to have a positive effect on primary mouse osteoblasts *in vitro* and in the form of cerium oxide nanoparticles acts as a neuroprotective agent.[83] The inclusion of gallium (Ga^{3+}), which increases bone calcium content, shows antimicrobial activity and inhibits osteoclast activity, hence improving the mechanical properties, has also been studied.[84] Studies concerning the development of cerium, gallium and zinc-containing sol-gel bioactive glasses have shown promising results for the delivery of these ions.[82–88]

5.5.2 Delivery of Drugs and Growth Factors from MBGs

Functional drugs and growth factors play an important role in inducing an active effect on bone regeneration and in combating infection of injuries and surgical wounds.[89,90] Previously, several polymeric materials have been employed as DDS because they are relatively biocompatible and biodegradable and they can encapsulate widely different types of drug molecules. However, it is not possible to obtain zero burst release with these materials because as soon as they are in contact with the biological fluids they start degrading and subsequently leak the encapsulated drugs (refer to Chapter 1). Moreover, the kinetic release is based on their degradation rate. For these reasons, research has more recently focussed on developing structurally stable materials as an alternative to soft degradable polymers.[9] MBGs fulfil all the requirements to be employed as DDS: good biocompatibility, high pore volume, high specific surface area, well-ordered mesoporous structures with tuneable pore dimension and high concentration of Si–OH groups that allows their functionalization.[9,11,21] MBGs have been used for delivering drugs, for example, dexamethasone[11,90] (DEX) and dimethyloxalylglycine[89] (DMOG), which can enhance osteogenesis for bone tissue regeneration.[91] DMOG in MBG scaffolds can mimic a hypoxic environment, which not only improves the angiogenic capability of hBMSCs, but also enhances their osteogenic differentiation.[89] Wu et al.[89] found that MBG scaffolds with slow DMOG release significantly increase the expression of bone-related genes more than those with instant DMOG release. DEX has been loaded in MBG as an osteogenic drug.[11,90] The release of DEX has been found to significantly enhance the bone-relative gene expression of osteoblasts.[90] MBGs can also be used for the delivery of antibacterial and anti-inflammatory molecules, such as gentamicin, ibuprofen, tetracycline and ampicillin. A schematic representation of some of the different biomolecules that have been loaded into MBGs is reported in Figure 5.7.

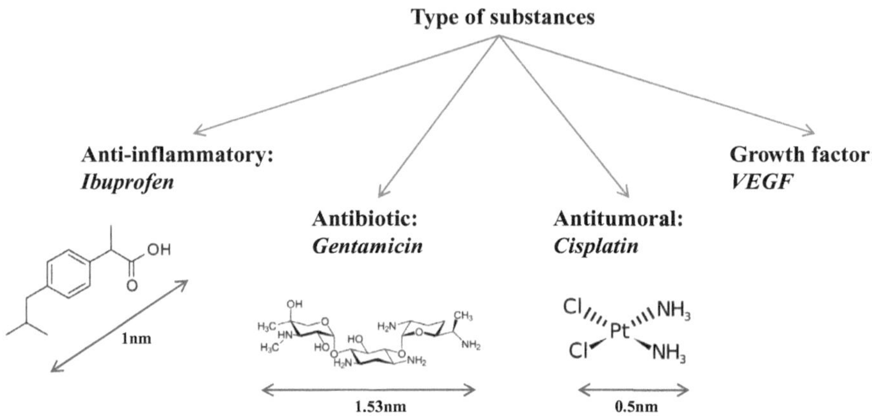

Figure 5.7 Schematic representation of the different biomolecules that have been loaded into MBGs for drug delivery applications.

A very straightforward technique for loading these biomolecules in a mesoporous structure was described by Xia and Chang.[12] Mesoporous 58S bioactive glass (M58S) and mesoporous 77S bioactive glass (M77S) were loaded with gentamicin sulphate (GS) by immersion into a gentamicin solution (10 mg/ml). For comparison, non-mesoporous 58S bioactive glass was also loaded with antibiotic molecules. The depletion method confirmed the successful loading procedure. MBG showed improved drug loading capability compared to non-MBG. Additionally, a correlation between the efficiency of drug loading and pore structure and volume was found. Depending on the pore diameter[92] and the hosting mechanism inside the mesopores, the drug release pattern was influenced. Figure 5.8 shows the different mechanisms that have been proposed for hosting a drug in MBG mesopores, as exemplified with GS.

Lin et al.[93] presented the functionalization of MBG with photoactive coumarin used as a drug delivery carrier and investigated drug storage/release characteristics using phenanthrene as a model drug. They reported the photo-controlled storage and release of guest molecules in coumarin-modified MBG, which was studied through the photo-responsive reversible intermolecular dimerization of coumarin derivatives attached preferentially to the pore outlets and the development of systems for releasing guest molecules from MBG.[93]

Growth factors can also be loaded into MBGs for controllable release in BTE for enhancing angiogenesis and osteogenesis.[76,94,95] Angiogenesis is

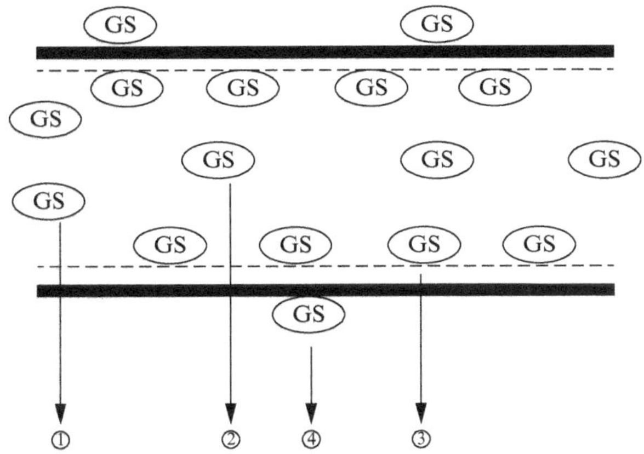

Figure 5.8 Gentamicin sulphate (GS) molecules hosted in MBG mesopores. (1) GS at the wall of mesopores, (2) GS inside mesopores without bonding, (3) GS inside mesopores with Si–OH and P–OH groups bonded *via* hydrogen bonds and (4) adsorbed GS molecule on the external MBG surface. *(Reproduced from Ref. 12 with permission of Elsevier Limited).*

directed by different growth factors in a complex multistep process. Vascular endothelial growth factor (VEGF) has been identified as a key active element; it is able to activate endothelial cells in the surrounding tissue by stimulating their migration, proliferation and the formation of tubular structures. For this reason VEGF has been loaded in MBG scaffolds, where the ordered mesostructure plays an important role in improving the loading efficiency, decreasing the burst release and maintaining the bioactivity of VEGF.[76,94,95] Also, osteogenic growth factors such as rhBMP-2 can be efficiently loaded into MBGs. These studies show that the system rhBMP-2/MBG may be clinically useful for bone regeneration of large defects.[96,97] MBGs have also been considered for use as a release system for curcumin, considered for its medicinal benefits with no side effects.[85] Shruti et al. reported that low Ce, Ga and Zn-substituted MBGs can be considered as drug release systems due to their high drug release, high bioactivity and added functional benefits of the incorporated therapeutic ions.[85]

5.5.3 Delivery of Drugs from a Polymer Coating of MBG

Drug delivery capability can also be introduced to MBGs by a polymer coating or alternatively, MBGs can be used to coat a substrate's surface to introduce drug delivery capabilities to the material. Ehlert et al.[68] accomplished bioactive coatings for medical implants using a coating of mesoporous silica for controlled drug delivery. The antibacterial drug ciprofloxacin was loaded onto the sulphonated mesoporous silica coating and the drug release kinetics was investigated. This study showed that by applying additional layers of bis (trimethylsilyl)hexane or dioctyltetramethyldisilazane, the release was delayed and the effective antibacterial activity period was prolonged.[68] Moreover, cell culture experiments demonstrated biocompatibility of the materials.[68] These results suggest that this is a promising approach to prevent bacterial implant infections during the most critical time period after implantation.

Another approach proposed by Bhattacharyya et al.[98] is the coating of a loaded mesoporous material with a polymer for the controlled release of macromolecules. Mesoporous silica nanoparticles incorporating a model protein molecule for growth factors (trypsin inhibitor) were coated with poly(ethylene glycol).[98] Sustained release of the encapsulated protein was demonstrated for a period of several weeks from this composite system, providing a basis for using composite nanoparticles for the controlled release of bioactive molecules under clinically relevant conditions.[98] Mesoporous materials offer a means of adsorbing and desorbing a range of biologically active substances, such as protein and growth factors, without the loss of conformation and biological function.[43,99] In addition, the texture of sol-gel materials can be varied through different processing conditions. Thus, dissolution rates and surface chemical binding sites of sol-gel products can be controlled which is important for the application of MBGs as delivery devices or as vehicles for eliciting specific *in vitro* cell culture behaviour.[43]

5.6 CONCLUSIONS AND FUTURE CONTEXT

Mesoporous materials are being increasingly investigated for a range of biomedical applications. As was highlighted, MBGs are good inorganic biomaterials to be used for controlled drug release, therapeutic ion release and growth factor release for bone regeneration. This is mainly due to their versatility in formulation and the ability to design the delivery system down to the nanostructural level. This is a new area of biomedical materials research which has great potential to be further developed for improved healthcare applications.

REFERENCES

1. Wu C, Chang J. Mesoporous bioactive glasses: structure characteristics, drug/growth factor delivery and bone regeneration application. *Interface Focus* 2012;**2**:292–306.
2. Misra SK, Ansari T, Mohn D, Valappil SP, Brunner TJ, Stark WJ, Roy I, Knowles JC, Sibbons PD, Jones EV, Boccaccini AR, Salih V. Effect of nanoparticulate bioactive glass

particles on bioactivity and cytocompatibility of poly(3-hydroxybutyrate) composites. *J R Soc Interface* 2010;**7**:453–65.

3. Hench LL, Thompson I. Twenty-first century challenges for biomaterials. *J R Soc* 2010;**7**:379–91.
4. Chen QZ, Thompson ID, Boccaccini AR. 45S5 Bioglass®-derived glass-ceramic scaffolds for bone tissue engineering. *Biomaterials* 2006;**27**:2414–25.
5. Jones JR, Tsigkou O, Coates EE, Stevens MM, Polak JM, Hench LL. Extracellular matrix formation and mineralization on a phosphate-free porous bioactive glass scaffold using primary human osteoblast (HOB) cells. *Biomaterials* 2007;**28**:1653–63.
6. Jones JR, Ehrenfried LM, Hench LL. Optimising bioactive glass scaffolds for bone tissue engineering. *Biomaterials* 2006;**27**:964–73.
7. Wu ZY, Hill RG, Yue S, Nightingale D, Lee PD, Jones JR. Melt-derived bioactive glass scaffolds produced by a gel-cast foaming technique. *Acta Biomater* 2011;**7**:1807–16.
8. Yan X, Yu C, Zhou X, Tang J, Zhao D. Highly ordered mesoporous bioactive glasses with superior *in vitro* bone-forming bioactivities. *Angew Chem Int Ed* 2004;**43**: 5980–4.
9. Vallet-Regí M, Manzano-García M, Colilla M. *Biomedical applications of mesoporous ceramics: drug delivery, smart materials and bone tissue engineering*. Boca Raton: CRC Press; 2012.
10. Zhao D, Wan Y, Zhou W. *Ordered mesoporous materials*. Weinheim: Wiley; 2013.
11. Wu C, Chang J. Multifunctional mesoporous bioactive glasses for effective delivery of therapeutic ions and drug/growth factors. *J Control Release* 2014;**193**:282–95.
12. Xia W, Chang J. Well-ordered mesoporous bioactive glasses (MBG): a promising bioactive drug delivery system. *J Control Release* 2006;**110**:522–30.
13. Vallet-Regí M, Izquierdo-Barba I, Colilla M. Structure and functionalization of mesoporous bioceramics for bone tissue regeneration and local drug delivery. *Philos Trans A Math Phys Eng Sci* 2012;**370**:1400–21.
14. Zhao D, Feng J, Huo Q, Melosh N, Fredrickson GH, Chmelka BF, Stucky GD. Triblock copolymer syntheses of mesoporous silica with periodic 50 to 300 Angstrom pores. *Science* 1998;**279**:548–52.
15. Chen H, Gu J, Shi J, Liu Z, Gao J, Ruan M, Yan D. A composite surfactant route for the synthesis of thermally stable and hierarchically porous zirconia with a nanocrystallized framework. *Adv Mater* 2005;**17**:2010–4.
16. López-Noriega A, Arcos D, Izquierdo-Barba I, Sakamoto Y, Terasaki O, Vallet-Regi M. Ordered mesoporous bioactive glasses for bone tissue regeneration. *Chem Mater* 2006;**18**:3137–44.
17. Zhao D, Yang P, Chmelka BF, Stucky GD. Multiphase assembly of mesoporous-macroporous membranes. *Chem Mater* 1999;**11**:1174–8.
18. Dong BA, Wang Y, Tang Y, Ren N, Zhang Y, Yue Y, Gao Z. Zeolitic tissue through wood cell templating. *Adv Mater* 2002;**14**:926–9.
19. Vallet-Regí M, Ruiz-González L, Izquierdo-Barba I, González-Calbet JM. Revisiting silica based ordered mesoporous materials: medical applications. *J Mater Chem* 2006;**16**:26.
20. Hartmann M. Ordered mesoporous materials for bioadsorption and biocatalysis. *Chem Mater* 2005;**17**:4577–93.
21. Vallet-Regí M. Ordered mesoporous materials in the context of drug delivery systems and bone tissue engineering. *Chem Eur J* 2006;**12**:5934–43.
22. Izquierdo-Barba I, Ruiz-González L, Doadrio JC, González-Calbet JM, Vallet-Regí M. Tissue regeneration: a new property of mesoporous materials. *Solid State Sci* 2005;**7**: 983–9.
23. Brinker CJ, Lu Y, Sellinger A, Fan H. Evaporation-induced self-assembly: nanostructures made easy. *Adv Mater* 1999;**11**:579–85.

24. Yan XX, Deng HX, Huang XH, Lu GQ, Qiao SZ, Zhao DY, Yu CZ. Mesoporous bioactive glasses. I. Synthesis and structural characterization. *J Non-Cryst Solids* 2005;**351**:3209–17.
25. Yan X, Huang X, Yu C, Deng H, Wang Y, Zhang Z, Qiao S, Lu G, Zhao D. The *in vitro* bioactivity of mesoporous bioactive glasses. *Biomaterials* 2006;**27**:3396–403.
26. Shi QH, Wang JF, Zhang JP, Fan J, Stucky GD. Rapid-setting, mesoporous, bioactive glass cements that induce accelerated *in vitro* apatite formation. *Adv Mater* 2006;**18**:1038–42.
27. Hench LL, Polak JM. Third-generation biomedical materials. *Science* 2002;**295**:1014–7.
28. Langer R, Vacanti JP. Tissue engineering. *Science* 1993;**260**:920–6.
29. Vallet-Regí M, Izquierdo-Barba I, Rámila A, Pérez-Pariente J, Babonneau F, González-Calbet JM. Phosphorous-doped MCM-41 as bioactive material. *Solid State Sci* 2005;**7**:233–7.
30. Salonen J, Laitinen L, Kaukonen AM, Tuura J, Björkqvist M, Heikkilä T, Vähä-Heikkilä K, Hirvonen J, Lehto VP. Mesoporous silicon microparticles for oral drug delivery: loading and release of five model drugs. *J Control Release* 2005;**108**:362–74.
31. Izquierdo-Barba I, Martinez A, Doadrio AL, Pérez-Pariente J, Vallet-Regí M. Release evaluation of drugs from ordered three-dimensional silica structures. *Eur J Pharm Sci Off J Eur Fed Pharm Sci* 2005;**26**:365–73.
32. Yang Q, Wang S, Fan P, Wang L, Di Y, Lin K, Xiao FS. pH-responsive carrier system based on carboxylic acid modified mesoporous silica and polyelectrolyte for drug delivery. *Chem Mater* 2005;**17**: 5999–6003.
33. Yun H, Kim S, Hyun Y, Heo S, Shin J. Hierarchically mesoporous-macroporous bioactive glasses scaffolds for bone tissue regeneration. *J Biomed Mater Res B Appl Biomater* 2008;**87**:374–80.
34. Yun H, Kim S, Hyeon Y. Highly ordered mesoporous bioactive glasses with Im3m symmetry. *Mater Lett* 2007;**61**:4569–72.
35. Vallet-Regí M. Revisiting ceramics for medical applications. *Dalton Trans* 2006;**44**:5211–20.
36. Manzano M, Vallet-Regí M. New developments in ordered mesoporous materials for drug delivery. *J Mater Chem* 2010;**20**:5593.
37. Vallet-Regí M, Balas F, Arcos D. Mesoporous materials for drug delivery. *Angew Chem Int Ed Engl* 2007;**46**:7548–58.
38. Horcajada P, Rámila A, Boulahya K, González-Calbet J, Vallet-Regí M. Bioactivity in ordered mesoporous materials. *Solid State Sci* 2004;**6**:1295–300.
39. Mouriño V, Boccaccini AR. Bone tissue engineering therapeutics: controlled drug delivery in three-dimensional scaffolds. *J R Soc* 2010;**7**:209–27.
40. Soler-Illia GJAA, Sanchez C, Lebeau B, Palarin J. Chemical strategies to design textured materials: from microporous and mesoporous oxides to nanonetworks and hierarchical structures. *Chem Rev* 2002;**102**:4093–138.
41. Polak JM, Hench LL, Kemp P. *Future strategies for tissue and organ replacement.* London: World Scientific; 2002.
42. Li R, Hench LL. An investigation of bioactive glass powders by sol-gel processing. *J Appl Biomater* 1991;**2**:231–9.
43. Sepulveda P, Jones JR, Hench LL. Bioactive sol-gel foams for tissue repair. *J Biomed Mater Res* 2002;**59**:340–8.
44. Ravaglioli A, Krajewski A. *Bioceramics: materials, properties, applications.* Dordrecht: Chapman & Hall; 1992.
45. Pereira MM, Clark AE, Hench LL. Effect of texture on the rate of hydroxyapatite formation on gel-silica surface. *J Am Ceram Soc* 1995;**78**:2463–8.
46. Pereira MM, Hench LL. Mechanisms of hydroxyapatite formation on porous gel-silica substrates. *J Sol-Gel Sci Technol* 1996;**7**:59–68.

47. Jones JR. New trends in bioactive scaffolds: the importance of nanostructure. *J Eur Ceram Soc* 2009;**29**:1275–81.
48. Zhu Y, Wu C, Ramaswamy Y, Kockrick E, Simon P, Kaskel S, Zreiqat H. Preparation, characterization and in vitro bioactivity of mesoporous bioactive glasses (MBGs) scaffolds for bone tissue engineering. *Microporous Mesoporous Mater* 2008;**112**: 494–503.
49. Li X, Li X, Wang X, Chen H, Jiang P, Dong X, Shi J. Hierarchically porous bioactive glass scaffolds synthesized with a PUF and P123 cotemplated approach. *Chem Mater* 2007;**19**:4322–6.
50. Lin H, Ma J, Li X, Wu X, Qu F. A co-templated approach to hierarchically mesoporous–macroporous bioactive glasses (MMBG) scaffolds for bone tissue regeneration. *J Sol-Gel Sci Technol* 2012;**62**:170–6.
51. Zhu Y, Zhang Y, Wu C, Fang Y, Yang J, Wang S. The effect of zirconium incorporation on the physiochemical and biological properties of mesoporous bioactive glasses scaffolds. *Microporous Mesoporous Mater* 2011;**143**:311–9.
52. Wu C, Miron R, Sculean A, Kaskel S, Doert T, Schulze R, Zhang Y. Proliferation, differentiation and gene expression of osteoblasts in boron-containing associated with dexamethasone deliver from mesoporous bioactive glass scaffolds. *Biomaterials* 2011; **32**:7068–78.
53. Wu C, Fan W, Gelinsky M, Xiao Y, Simon P, Schulze R, Doert T, Luo Y, Cuniberti G. Bioactive SrO-SiO_2 glass with well-ordered mesopores: characterization, physiochemistry and biological properties. *Acta Biomater* 2011;**7**:1797–806.
54. Wu C, Fan W, Zhu Y, Gelinsky M, Chang J, Cuniberti G, Albrecht V, Friis T, Xiao Y. Multifunctional magnetic mesoporous bioactive glass scaffolds with a hierarchical pore structure. *Acta Biomater* 2011;**7**:3563–72.
55. Wang H, Gao X, Wang Y, Tang J, Sun C, Deng X, Niu X. Bio-templated synthesis of mesoporous bioactive glass with a hierarchical pore structure. *Mater Lett* 2012;**76**: 237–9.
56. Yun H, Kim S, Hyun Y, Heo S, Shin J. Three-dimensional mesoporous–giantporous inorganic/organic composite scaffolds for tissue engineering. *Chem Mater* 2007;**19**:6363–6.
57. García A, Izquierdo-Barba I, Colilla M, de Laorden CL, Vallet-Regí M. Preparation of 3-D scaffolds in the SiO_2-P_2O_5 system with tailored hierarchical meso-macroporosity. *Acta Biomater* 2011;**7**:1265–73.
58. Franco J, Hunger P, Launey ME, Tomsia AP, Saiz E. Direct write assembly of calcium phosphate scaffolds using a water-based hydrogel. *Acta Biomater* 2010;**6**:218–28.
59. Miranda P, Pajares A, Saiz E, Tomsia AP, Guiberteau F. Mechanical properties of calcium phosphate scaffolds fabricated by robocasting. *J Biomed Mater Res* 2008;**A85**: 218–27.
60. Miranda P, Saiz E, Gryn K, Tomsia AP. Sintering and robocasting of beta-tricalcium phosphate scaffolds for orthopaedic applications. *Acta Biomater* 2006;**2**:457–66.
61. Wu C, Luo Y, Cuniberti G, Xiao Y, Gelinsky M. Three-dimensional printing of hierarchical and tough mesoporous bioactive glass scaffolds with a controllable pore architecture, excellent mechanical strength and mineralization ability. *Acta Biomater* 2011;**7**:2644–50.
62. Slowing II, Vivero-Escoto JL, Wu C-W, Lin VS-Y. Mesoporous silica nanoparticles as controlled release drug delivery and gene transfection carriers. *Adv Drug Deliv Rev* 2008;**60**:1278–88.
63. Monsigny M, Roche AC, Midoux P, Mayer R. Glycoconjugates as carriers for specific delivery of therapeutic drugs and genes. *Adv Drug Deliv Rev* 1994;**14**:1–24.
64. Hoffmann F, Cornelius M, Morell J, Fröba M. Silica-based mesoporous organic-inorganic hybrid materials. *Angew Chem Int Ed Engl* 2006;**45**:3216–51.
65. Stevens MM. Biomaterials for bone tissue engineering. *Mater Today* 2008;**11**:18–25.

66. Zhu Y, Kaskel S. Comparison of the in vitro bioactivity and drug release property of mesoporous bioactive glasses (MBGs) and bioactive glasses (BGs) scaffolds. *Microporous Mesoporous Mater* 2009;**118**:176–82.
67. Hum J, Boccaccini AR. Bioactive glasses as carriers for bioactive molecules and therapeutic drugs: a review. *J Mater Sci Mater Med* 2012;**23**:2317–33.
68. Ehlert N, Badar M, Christel A, Lohmeier SJ, Luessenhop T, Stieve M, Lenarz T, Mueller PP, Behrens P. Mesoporous silica coatings for controlled release of the antibiotic ciprofloxacin from implants. *J Mater Chem* 2011;**21**:752.
69. Hoppe A, Güldal NS, Boccaccini AR. A review of the biological response to ionic dissolution products from bioactive glasses and glass-ceramics. *Biomaterials* 2011;**32**:2757–74.
70. Wu C, Zhou Y, Xu M, Han P, Chen L, Chang J, Xiao Y. Copper-containing mesoporous bioactive glass scaffolds with multifunctional properties of angiogenesis capacity, osteostimulation and antibacterial activity. *Biomaterials* 2013;**34**:422–33.
71. Gargiulo N, Cusano AM, Causa F, Caputo D, Netti PA. Silver-containing mesoporous bioactive glass with improved antibacterial properties. *J Mater Sci Mater Med* 2013;**24**:2129–35.
72. Zhang Y, Wei L, Chang J, Miron RJ, Shi B, Yi S, Wu C. Strontium-incorporated mesoporous bioactive glass scaffolds stimulating in vitro proliferation and differentiation of bone marrow stromal cells and in vivo regeneration of osteoporotic bone defects. *J Mater Chem* 2013;**B1**:5711.
73. Zhu Y, Li X, Yang J, Wang S, Gao H, Hanagata N. Composition-structure-property relationships of the CaO-M_xO_y-SiO_2-P_2O_5 (M = Zr, Mg, Sr) mesoporous bioactive glass (MBG) scaffolds. *J Mater Chem* 2011;**21**:9208.
74. Wu C, Zhou Y, Lin C, Chang J, Xiao Y. Strontium-containing mesoporous bioactive glass scaffolds with improved osteogenic/cementogenic differentiation of periodontal ligament cells for periodontal tissue engineering. *Acta Biomater* 2012;**8**:3805–15.
75. Fan W, Crawford R, Xiao Y. Enhancing in vivo vascularized bone formation by cobalt chloride-treated bone marrow stromal cells in a tissue engineered periosteum model. *Biomaterials* 2010;**31**:3580–9.
76. Lode A, Wolf-Brandstetter C, Reinstorf A, Bernhardt A, König U, Pompe W, Gelinsky M. Calcium phosphate bone cements, functionalized with VEGF: release kinetics and biological activity. *J Biomed Mater Res Part* 2007;**A81A**:474–83.
77. Fu O-Y, Hou M-F, Yang S-F, Huang S-C, Lee W-Y. Cobalt chloride-induced hypoxia modulates the invasive potential and matrix metalloproteinases of primary and metastatic breast cancer cells. *Anticancer Res* 2009;**29**:3131–8.
78. Han P, Wu C, Chang J, Xiao Y. The cementogenic differentiation of periodontal ligament cells via the activation of Wnt/β-catenin signalling pathway by Li^+ ions released from bioactive scaffolds. *Biomaterials* 2012;**33**:6370–9.
79. Zhai W, Lu H, Wu C, Chen L, Lin X, Naoki K, Chen G, Chang J. Stimulatory effects of the ionic products from Ca-Mg-Si bioceramics on both osteogenesis and angiogenesis in vitro. *Acta Biomater* 2013;**9**:8004–14.
80. Wu C, Zhou Y, Fan W, Han P, Chang J, Yuen J, Zhang M, Xiao Y. Hypoxia-mimicking mesoporous bioactive glass scaffolds with controllable cobalt ion release for bone tissue engineering. *Biomaterials* 2012;**33**:2076–85.
81. Clément-Lacroix P, Ai M, Morvan F, Roman-Roman S, Vayssière B, Belleville C, Estrera K, Warman ML, Baron R, Rawadi G. Lrp5-independent activation of Wnt signaling by lithium chloride increases bone formation and bone mass in mice. *Proc Natl Acad Sci U S A* 2005;**102**:17406–11.
82. Sánchez-Salcedo S, Shruti S, Salinas AJ, Malavasi G, Menabue L, Vallet-Regí M. In vitro antibacterial capacity and cytocompatibility of SiO_2-CaO-P_2O_5 meso-macroporous glass scaffolds enriched with ZnO. *J Mater Chem* 2014;**B2**:4836–47.

83. Shruti S, Salinas AJ, Lusvardi G, Malavasi G, Menabue L, Vallet-Regi M. Mesoporous bioactive scaffolds prepared with cerium-, gallium- and zinc-containing glasses. *Acta Biomater* 2013;**9**:4836–44.
84. Aina V, Morterra C. Ga-modified (Si-Ca-P) sol-gel glasses: possible relationships between surface chemical properties and bioactivity. *J Phys Chem* 2011;**C115**:22461–74.
85. Shruti S, Salinas AJ, Ferrari E, Malavasi G, Lusvardi G, Doadrio AL, Menabue L, Vallet-Regi M. Curcumin release from cerium, gallium and zinc containing mesoporous bioactive glasses. *Microporous Mesoporous Mater* 2013;**180**:92–101.
86. Lusvardi G, Malavasi G, Menabue L, Shruti S. Gallium-containing phosphosilicate glasses: functionalization and in vitro bioactivity. *Mater Sci Eng C Mater Biol Appl* 2013;**33**:3190–6.
87. Shruti S, Salinas AJ, Malavasi G, Lusvardi G, Menabue L, Ferrara C, Mustarelli P, Vallet-Regí M. Structural and in vitro study of cerium, gallium and zinc containing sol–gel bioactive glasses. *J Mater Chem* 2012;**22**:13698.
88. Salinas AJ, Shruti S, Malavasi G, Menabue L, Vallet-Regí M. Substitutions of cerium, gallium and zinc in ordered mesoporous bioactive glasses. *Acta Biomater* 2011;**7**:3452–8.
89. Wu C, Zhou Y, Chang J, Xiao Y. Delivery of dimethyloxallyl glycine in mesoporous bioactive glass scaffolds to improve angiogenesis and osteogenesis of human bone marrow stromal cells. *Acta Biomater* 2013;**9**:9159–68.
90. Wu C, Fan W, Chang J, Xiao Y. Mussel-inspired porous SiO_2 scaffolds with improved mineralization and cytocompatibility for drug delivery and bone tissue engineering. *J Mater Chem* 2011;**21**:18300.
91. Wu C, Fan W, Chang J. Functional mesoporous bioactive glass nanospheres: synthesis, high loading efficiency, controllable delivery of doxorubicin and inhibitory effect on bone cancer cells. *J Mater Chem* 2013;**B1**:2710.
92. Xue JM, Shi M. PLGA/mesoporous silica hybrid structure for controlled drug release. *J Control Release* 2004;**98**:209–17.
93. Lin H-M, Wang W-K, Hsiung P, Shyu S-G. Light-sensitive intelligent drug delivery systems of coumarin-modified mesoporous bioactive glass. *Acta Biomater* 2010;**6**: 3256–63.
94. Wu C, Fan W, Chang J, Xiao Y. Mesoporous bioactive glass scaffolds for efficient delivery of vascular endothelial growth factor. *J Biomater Appl* 2013;**28**:367–74.
95. Oladipupo S, Hu S, Kovalski J, Yao J, Santeford A, Sohn RE, Shohet R, Maslov K, Wang LV, Arbeit JM. VEGF is essential for hypoxia-inducible factor-mediated neovascularization but dispensable for endothelial sprouting. *Proc Natl Acad Sci U S A* 2011;**108**:13264–9.
96. Nash AD, Baca M, Wright C, Scotney PD. The biology of vascular endothelial growth factor-B (VEGF-B). *Pulm Pharmacol Ther* 2006;**19**:61–9.
97. Soker S, Machado M, Atala A. Systems for therapeutic angiogenesis in tissue engineering. *World J Urol* 2000;**18**:10–8.
98. Bhattacharyya S, Wang H, Ducheyne P. Polymer-coated mesoporous silica nanoparticles for the controlled release of macromolecules. *Acta Biomater* 2012;**8**:3429–35.
99. Hench LL. Sol-gel materials for bioceramic applications. *Curr Opin Solid State Mater Sci* 1997;**2**:604–10.

CHAPTER 6

Calcium Phosphate Materials for Controlled Release Systems

Contents

6.1 Calcium Phosphate Materials and Their Properties 161
6.2 Calcium Phosphate Cements 164
6.2.1 The fundamentals of calcium phosphate cement chemistry 165
6.2.2 Calcium phosphate cements as drug delivery systems 169
6.2.3 Introducing drugs into calcium phosphate cement systems 170
6.2.4 Controlling drug release rates 171
6.2.5 Calcium phosphate cements—additional functionality 176
6.3 Calcium Phosphate Foams for Drug Delivery 178
6.4 Drug Release from Calcium Phosphate Coatings 179
References 183

6.1 CALCIUM PHOSPHATE MATERIALS AND THEIR PROPERTIES

Calcium and orthophosphate ions are two of the most biologically important inorganic species known to the medical world. Calcium ions have vital roles within neuron signalling pathways, controlling muscle contraction/relaxation, as co-factors for many enzymes and for maintaining potential difference across cell membranes.[1] The many biological functions of phosphate ions are even more varied. For example, the free energy released during the hydrolysis of adenosine triphosphate (ATP) to adenosine diphosphate (ADP) and subsequently to adenosine monophosphate provides the energy to power most of the biochemical reactions vital to life. Phosphate groups also make up the backbone of DNA and RNA. In a nod to the versatility of the biochemical landscape, and the wide-ranging physicochemical properties of the phosphate group, one of its other most important functions is as a major constituent of the structural components of the body—bones and teeth. In comparison to the covalently bound phosphate units in ATP, hydrolysis of which to yield ADP is a highly favourable process and occurs spontaneously in water, crystalline, ionic calcium phosphates, the principal inorganic component of bones, are exceptionally insoluble in water (Table 6.1 and Figure 6.1).

Inorganic Controlled Release Technology
http://dx.doi.org/10.1016/B978-0-08-099991-3.00006-5

Table 6.1 Properties of major calcium phosphate phases

Calcium phosphate	Abbreviation	Chemical formula	Ca/P ratio	Solubility product (K_{sp})	Solubility at 37 °C (mg L^{-1})
Monocalcium phosphate monohydrate	MCPM	$Ca(H_2PO_4)_2 \cdot H_2O$	0.5	7.2×10^{-2}	
Monocalcium phosphate	MCP	$Ca(H_2PO_4)_2$	0.5	7.2×10^{-2}	
Dicalcium phosphate dihydrate	DCPD	$CaHPO_4 \cdot 2H_2O$	1.0	2.5×10^{-7}	74
Dicalcium phosphate	DCP	$CaHPO_4$	1.0	1.26×0^{-7}	41
Octacalcium phosphate	OCP	$Ca_8H_2(PO_4)_6 \cdot 5H_2O$	1.33	2.51×10^{-97}	1.8×10^{-2}
α-Tricalcium phosphate	α-TCP	$Ca_3(PO_4)_2$	1.5	3.16×10^{-26}	0.24
β-Tricalcium phosphate	β-TCP	$Ca_3(PO_4)_2$	1.5	1.25×10^{-29}	0.15
Amorphous calcium phosphate	ACP	$Ca_3(PO_4)_2 \cdot nH_2O$	1.2-2.2	Varies due to being metastable	
Calcium deficient hydroxyapatite	CDHA	$Ca_{10-x}(HPO_4) \times (PO_4)_{6-x}(OH)_{2-x}$ $(0 < \times < 1)$	1.5-1.67	Dependent on stoichiometry	
Carbonated apatite	CA	$Ca_5(PO_4,CO_3)_3$	1.67		
Hydroxyapatite	HA	$Ca_{10}(PO_4)_6(OH)_2$	1.67	2.35×10^{-59}	9.6×10^{-5}
Oxyapatite	OXA	$Ca_{10}(PO_4)_6O$	1.67		
Tetracalcium phosphate	TTCP	$Ca_4(PO_4)_2O$	2.0	1×10^{-38}-1×10^{-44}	0.39-0.075

Adapted from Refs. 2,3.

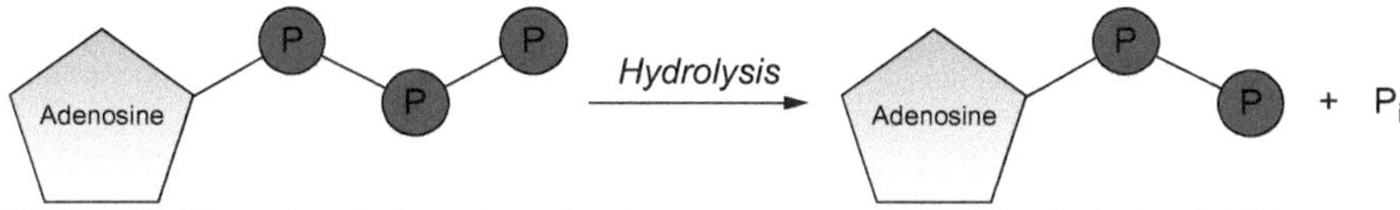

Figure 6.1 The role of phosphate in the energy generating hydrolysis of ATP.

Variations in the calcium:phosphate ratio, their crystallinity and their phase purity are the three most important factors in determining the bioactivity of calcium phosphate materials. As seen in Table 6.1, a number of distinct calcium phosphate phases exist largely defined by their calcium:phosphate ratios.

All calcium phosphate materials are highly insoluble at neutral pH, but their solubility increases under more acidic conditions. Anhydrous and hydrated monocalcium phosphates are not biocompatible as they are highly acidic and the generation of a local low pH environment accelerates their solubility dramatically.

Calcium deficient hydroxyapatite (CDHA) is most similar to bone, except that bone typically contains 7.4 wt% carbonate ions.[4] Hydroxyapatite (HA) and β-tricalcium phosphate (β-TCP) are the most commonly used calcium phosphates because of their osteogenic nature and their enhanced ability to bond with bone tissues. β-TCP is much more soluble than HA, and a common method of tailoring the solubility of calcium phosphate products is by preparing biphasic calcium phosphates with differing levels of HA and β-TCP, the higher the β-TCP amount the higher the solubility (Table 6.2).[5]

Table 6.2 The relative compositions of HA and natural bone

Component	Bone	Hydroxyapatite
Calcium (wt%)	34.8	39.6
Phosphorus (wt%)	15.2	18.5
Ca/P (molar ratio)	1.71	1.67
Sodium (wt%)	0.9	-
Magnesium (wt%)	0.72	-
Potassium (wt%)	0.03	-
Carbonate (wt%)	7.4	-
Fluoride (wt%)	0.03	-
Chloride (wt%)	0.13	-
Pyrophosphate (wt%)	0.07	-
Total inorganic (wt%)	65	100
Total organic (wt%)	25	-
Water (wt%)	10	-

Adapted from Ref. 2.

The chemistry of calcium phosphate materials—their preparation, interconversions, stabilities, degradation products, etc.—is well known and beyond the scope of discussion here. The reader is referred to the appropriate reference texts/papers.[3,6–12]

Calcium phosphates are attractive biomaterials because of the range of their potential bioactivity and biodegradability. These properties indicate that calcium phosphate-based materials are important components in many new biomedical materials and, in particular, in the use of medical devices.

6.2 CALCIUM PHOSPHATE CEMENTS

A most interesting and detailed review by Dorozhkin[13] reveals that an exogenous calcium phosphate was first used for the purpose of filling a lab-produced bone void in as early as 1920. The addition of a TCP slurry to deionized water demonstrated accelerated bone regrowth as compared with a control material.[14] The use of calcium phosphate materials as dental hygiene products was first reported in 1945,[15] and it soon followed that varying mixtures of calcium and phosphorus sources could have self-setting properties.[16] Köster was arguably the first to use self-setting calcium phosphates or calcium phosphate cements in a biomedical setting.[17] Porous implants were first prepared from different mixtures of tetra-, tri- and di-calcium phosphates with monocalcium phosphate. However, it was in the early 1980s when the benefits of self-setting calcium phosphate cements were realized and their use in treating dental caries was first established.[18–20]

In the context of an inorganic biomaterial for the treatment of bone voids or defects, the main benefits of calcium phosphate cements are their short hardening times and the relatively low level of heat generation which occurs when the cement is applied *in vivo* after mixing. Unlike the application of acrylate-based cements, the setting reactions are not exothermic preventing local tissue damage and allowing the inclusion of heat-sensitive organic molecules.[21] A viscous mouldable paste is initially formed which can be hand moulded or in some cases injected. Other advantages for bone applications are high bioactivity, the ability to bond directly to bone tissue and their osteoconductivity (Table 6.3).

As will be highlighted below, researchers are now developing methods and approaches to designing the bioresorption rate through careful control of the cement properties, in particular the composition and microstructural

Table 6.3 Comparison of PMMA cement setting temperatures vs. some commercially available calcium phosphate cements

Type	Code/brand	*T* (°C)	*t* (h)	*P* (min)	*s* (%)
PMMA	Rostal	73±3	13±1	13±2	1.05±0.07
	CMW-1	85±5	17±1	22±3	0 85±0.10
Calcium phosphate	Ml4	−20±10	0.2±0.1	-	0.10±0.05
	F36	Ambient	-	-	0 10±0.05
	R45	33±*2*	0.2±0.1	-	0.2±0.1
	H90	Ambient	-	-	0.10±0.05
	H72	Ambient	-	-	0.10±0.05
	V2	44±2	0.5±0.1	0.7±0.1	0 10±0.05

Ultimate temperature (*T*), time after the beginning of mixing (*t*) and period during which a cement was above body temperature (*p*) ($n=3$). Also the linear setting shrinkage is given (*s*).
Adapted with kind permission from Ref. 22 Springer Science and Business Media.

features. Excessive porosity in these materials, however, leads to poor mechanical strength which at this point makes them unsuitable for use in load-bearing applications. Other potential drawbacks for these new biomaterials include poor rheological properties leading to problems with injectability and also overly slow bioresorption rates. As is the case with the search for the perfect biomaterial for all application areas, the design space to choose from is huge, whilst the window of opportunity for the perfect materials solution is tiny in comparison. It is therefore vital to aim for the preparation of synthetic biomaterial platforms which are easily modifiable so that the release of important biomedical agents can be as broad as possible for the treatment of a range of conditions and disease states.

6.2.1 The Fundamentals of Calcium Phosphate Cement Chemistry

In a clinical setting, the nature of the way that calcium phosphate cements are formed means that the clinician applying the material must have the skill and experience to be able to handle, work, and apply the cement within the time frame of which the cement is 'workable'. In order to do this, the clinician must have absolute confidence that the cement components are going to mix and set consistently with every application. To guarantee this, the materials scientists, chemists and engineers involved with the design and manufacture of novel calcium phosphate cement materials must have a detailed knowledge of the chemical and physical processes involved during the cement setting process and the consequences that anticipated real-life variations might have.

Calcium phosphate cements are typically formed from the addition of an aqueous solution to a mixture of two different calcium phosphate phases, though there are many examples where three or more precursor phases are used. The setting process is a result of a series of dissolution and re-precipitation reactions and hardening is achieved through the random entanglement of the product crystals. Predominantly two product phases are achievable from these processes—HA or dicalcium phosphate dihydrate (DCPD)—depending on the pH of the setting system. HA forms preferentially at pH >4.2 and DCPD forms preferentially at pH <4.2. The formation of these two particular phases can be justified on thermodynamic grounds as these are the two most stable phases in the two particular pH ranges. The calcium phosphate cement-forming reactions can therefore be broken down into two basic processes: *apatite-forming* reactions and *brushite-forming* reactions.

Apatite products can be formed from single component systems—hydrolysis of α-tricalcium phosphate (α-TCP) powders to CDHA occurs without a change in the Ca/P ratio (1.5). The hydrolysis of α-TCP with water is slow so typically a source of phosphate ions is added to the liquid cement phase to enable a reasonable setting time.[23]

Apatite products can also be formed from multi-component systems of start materials. Combinations of two or more calcium phosphate precursors, some of which are more acidic and some of which are more basic, are employed. The basic component tends to be tetracalcium phosphate (TTCP) whilst the acidic component tends to be dicalcium phosphate (anhydrous and/or dihydrate). Acid-base reactions between the precursor materials result in dissolution and precipitation of the apatitic products, the Ca/P ratios of which are dependent on the relative ratios of the starting materials.

Brushite products are formed by multi-component precursor systems only and are generated by the acid-base dissolution and precipitation reactions of β-TCP, the basic component and monocalcium phosphate (anhydrous or monohydrate forms) or phosphoric acid, the acidic components.[24,25]

The cement setting reaction occurs *via* three distinct steps; dissolution of the precursor powders, which generates a local supersaturation of calcium and phosphate ions; nucleation of the new crystal phases around the precursor powder particles; further crystal growth as the dissolution of the precursor materials continues. In the initial stages, formation of the new crystal phases is dictated by the rate of dissolution of the start materials. As the cement setting reactions continue however, the rate determining process becomes diffusion of the start material ions through the newly formed layer of crystalline material.[26]

The microstructure of the newly formed crystal phases is fundamental in determining the properties of the resulting calcium phosphate materials. As new crystals are formed and grow, they interlock, which is primarily what contributes to the mechanical strength of the resulting products. This crystal interlocking process forms microporous voids, whilst the simultaneous formation of a macroporous network also occurs by the packing and association of larger crystalline aggregates (contrary to IUPAC definitions, in the area of calcium phosphate cements literature precedent suggests microporous voids are <100 nm. Macroporous voids are >100 μm).

The macroporous network has been found to be the most important in the context of using these materials in drug delivery applications and a further level of control over formation of the macropores can be generated by calcium phosphate foams (see Section 6.3). This intrinsic porosity makes these materials unsuitable for load-bearing applications but does offer benefits for their use in other applications. For example, bioactivity and bioresorption are increased as a result of the increased surface area, whilst their porous structure makes these products promising candidate materials for drug and active delivery. Pore dimensions, pore size distribution, specific surface area and interconnected porosity are all important physical properties and can be affected by the chosen processing conditions during the production of the cement products. It has been found that parameters such as the liquid:powder ratio of the cement precursors and the particle sizes of the precursor powders are, in particular, important factors for determining the product crystal morphology, and as an extension, the nature of the porous network (Figure 6.2).[27]

Reduction of the cement powder particle size results in a vastly increased surface area, which, in turn, increases the reactivity of the particle surfaces accordingly. This increased surface reactivity leads to faster dissolution kinetics and as such a higher degree of calcium and phosphate ion supersaturation within the local environment of the setting cement. This favours crystal nucleation and the precipitation of a greater number of smaller crystals with needle morphology. Conversely, when the particle surface area is larger, slower dissolution kinetics results in a lower degree of supersaturation, and the precipitation of a fewer number of larger plate-like crystals.

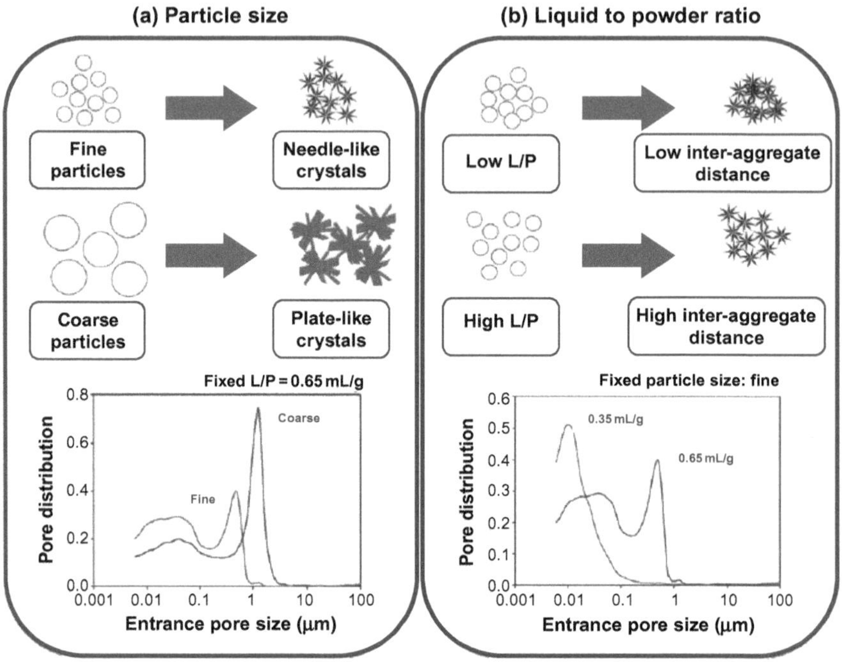

Figure 6.2 Control over calcium phosphate cement porosity shown schematically. *(Reprinted from Ref. 27 with permission from Elsevier).*

Plate-like crystals pack and aggregate much less efficiently as compared with the needle crystals which results in a larger pore size distribution, but importantly, a similar total pore volume overall (Figure 6.3).[28]

When lower liquid:powder ratios are used in the cement-forming reaction, the more concentrated suspension reduces the initial inter-particle distance. At the initiation of reaction, new calcium phosphate crystals are formed on the boundaries of the original precursor powder particles. At lower liquid:powder ratios then, as the suspension is more concentrated, a lower inter-aggregate distance at the onset of crystal formation results and thus leads to a pore size distribution dominated by smaller pores. When the liquid:powder ratio is higher, the suspension is more dilute and the precursor inter-particle distance is larger leading to a larger inter-aggregate distance at the onset of new crystal formation. This, in turn, leads to a porous network with larger average pore diameters and pore entrances (Figure 6.4).[27]

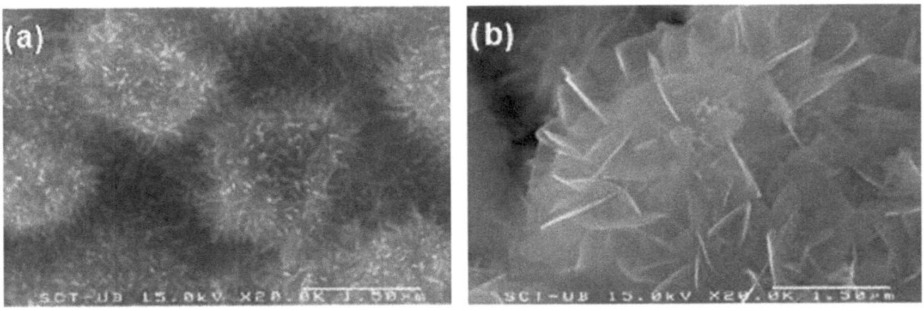

Figure 6.3 SEM images showing the relative differences in crystal morphology and pore size between: (a) smaller precursor particle size (needles) and (b) larger precursor particle size (plates). *(Reprinted from Ref. 28 with permission from Elsevier).*

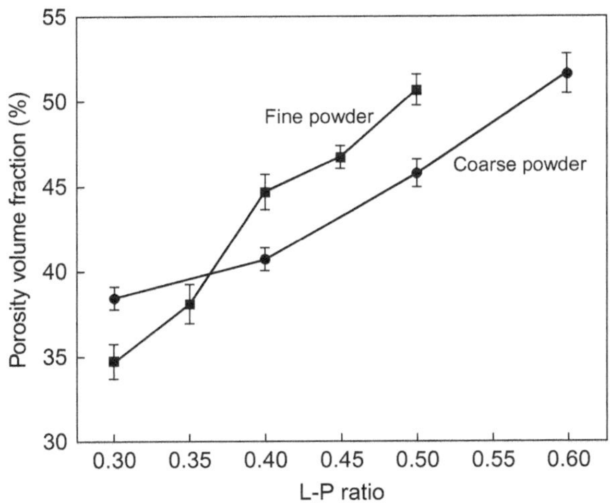

Figure 6.4 Increase in pore volume fraction with increased liquid:powder ratio for both fine and coarse cement precursor powders. *(Reprinted from Ref. 29 with permission from Elsevier).*

6.2.2 Calcium Phosphate Cements as Drug Delivery Systems

Calcium phosphate materials have been well studied as therapeutic biomaterials. Only more recently though have researchers attempted to manipulate the natural properties of these materials to make novel drug delivery systems. The inorganic calcium phosphate carrier system provides bioactivity, bioresorptivity and ideally osteoinductive behaviour, whilst the inclusion and associated release of antibiotics, anti-inflammatories, immunosuppressants and anti-cancer drug compounds makes these multi-functional materials promising candidates for controlled release drug delivery systems.

Research into the use of calcium phosphate cements as drug delivery systems for treating osseous defects has increased in recent years,[27] though most studies have focussed on the impact the addition of a drug compound to the cement-forming system has on the important physical properties associated with the formation of these materials; properties such as the setting time, the mechanical strength and the level of bioactivity. Only as research into the addition of drugs to these materials has developed are efforts being made to:

(i) elucidate the important cement properties affecting the drug release;
(ii) develop strategies for designing and controlling the drug release rate.

The dissolution of the cement powders, and the subsequent precipitation from supersaturated solutions of new calcium phosphate phases, involves complex reaction kinetics and as such slight variations in individual reaction methodologies and processes makes it difficult to draw cross-study comparisons. Nevertheless, excellent work by a number of different research groups has begun to yield promising, consistent results in elucidating some of the exploitable process parameters during the preparation of drug-loaded calcium phosphate cements.

6.2.3 Introducing Drugs into Calcium Phosphate Cement Systems

Drug compounds can be added to calcium phosphate cements by two fundamentally different approaches: the drug can be present during the cement-forming process or added to the as-formed calcium phosphate product. In the case where the drug is present during the cement-forming process, the drug can either be physically mixed with the cement powder precursors, it can be adsorbed to the surface of one of the cement powder precursors, or it can be added as a solution in the liquid phase of the cement-forming process. Typically, this is water or a dilute acidic solution—phosphoric acid, sodium phosphate and citric acid being common examples. The pH of this liquid phase has an important role in determining the specific calcium phosphate product phases and as such the inclusion of any drug compound which may have an impact on the pH of that solution is going to have an impact on the cement products formed. More recently, the use of organic liquids as the liquid phase of the cement-forming process has been researched. The principle behind this approach is that the ingress of bodily liquids into the cement formulation once applied *in vivo* is responsible for initiating and maintaining the formation of the calcium phosphate products. The organic liquid phase can either be based

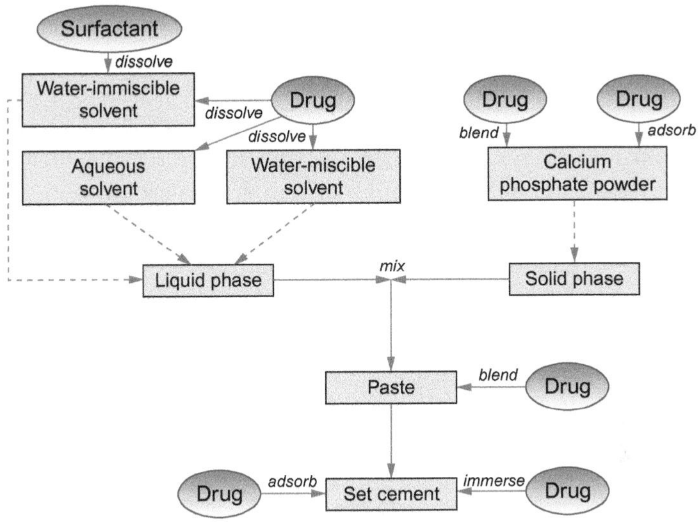

Figure 6.5 The different processing approaches to drug-loaded calcium phosphate cements.

on a water-miscible organic solvent such as glycerine[30,31] or polyethyleneglycol.[32] More recently, the use of water-immiscible solvents such as triglycerides[33,34] and paraffin oil[35] has been investigated, though these cement formulations typically have to contain a surfactant as an additional component to facilitate water diffusion through the cement-carrying solvent. These different approaches to the formation of drug-loaded calcium phosphate inorganic CRT systems are summarized in Figure 6.5.

An alternative, and arguably simpler, drug-loading method involves the pre-formation of the set calcium phosphate cement followed by adsorption of the drug from a concentrated drug solution onto the cement surface and into the porous network. This approach is fundamentally simpler in that the impact the presence of the drug is having on the formation of the developing calcium phosphate crystal network does not need to be considered. The formation of carrier systems with known and predictable porosity properties can therefore be carried out.

6.2.4 Controlling Drug Release Rates

Calcium phosphate cements can be considered to be non-swelling, matrix-type drug delivery systems. In addition, due to the very poor solubility and dissolution rates of calcium phosphate materials, the rate of carrier dissolution is very much slower than the rate of drug release. As a result, the drug release kinetics are generally controlled by diffusion mediated processes, with Fick's Laws of diffusion forming the basis for many of the models used to describe the release kinetics (see Chapter 2). Materials research has therefore focussed on ways in which the porosity of the product materials can be controlled. The following pore characteristics can all play significant roles in controlling drug release: pore volume, tortuosity (τ), pore size distribution, pore entrance size and mean path length to the surface.

For systems which are formed by an *in situ* drug-loading process, the drug can be considered to be almost completely located within the void spaces in between the crystal aggregates; it is extremely unlikely that the drug molecules are to be located within the crystallites themselves. For the post-loading method, the drug molecules are necessarily in the crystal structure void spaces since these are the only solvent accessible pores. Depending on the physicochemical properties of the drug and their method of loading, the drug molecules can be molecularly dispersed (within the pore fluid or adsorbed to the crystal surfaces) or form crystallites/aggregates themselves. The specific interactions between the drug and the calcium phosphate crystals at the molecular level are important—the rate of drug release is high unless the drug molecules are adsorbed to the crystal surfaces, as opposed to being in solvent-filled pores. Drug release from most cement products is dependent upon the rate of drug diffusion through the cement as well as on the solubility of the drug in question, as given by the Noyes-Whitney equation:

$$\frac{\mathrm{d}w}{\mathrm{d}t} = \frac{AD}{d}(C_s - C_b) \tag{6.1}$$

where ($\mathrm{d}w/\mathrm{d}t$) is the rate of drug dissolution, A is the surface area in contact with the dissolution medium, D is the diffusion coefficient, d is the thickness of the diffusion layer, C_s is the drug concentration in the diffusion layer and C_b is the drug concentration in the bulk dissolution medium.

Otsuka carried out a number of pioneering studies into active release from calcium phosphate cements. These studies were focussed on the use of different drug types and for different therapeutic areas rather than any systematic investigation of how to control or influence active release rate. This is understandable given that even now, as it was during those early studies, the primary function of calcium phosphate cement biomaterials is to act as osteoinductive, reinforcing bone defect fillers—the drug release is a secondary benefit of these new multi-functional materials.

Otsuka's cements are roughly based on Brown's pioneering work into calcium phosphate cements.[36] Typically, solid mixtures of TTCP, DCPD and precipitated HA are blended to give a homogenous mixture of different components.[37] The addition of HA as a component in this mixture is important for subsequently providing nucleating sites for new crystal growth.[38] This powder blend is then added to a dilute solution of phosphoric acid to form a paste. The drug compounds under investigation are then added to this paste and the formulations left to set. The loading and delivery of examples of antibiotics (norfloxacin,[37] cephalexin[37,39]), non-steroidal anti-inflammatories (indomethacin,[40] aspirin[41]), biomolecules (bovine serum albumin, bovine insulin[42]) and immunosuppressants (6-mercaptopurine[43,44]) were all investigated by Otsuka.

Some key findings came out of these early studies and have since been expanded on and investigated in thorough detail.[45] By showing that drug release could be fitted to the Higuchi kinetics (refer to Chapter 2 and Equations (6.2) and (6.3)) and as a result, confirming that drug release is diffusion controlled, this elucidated the fact that control over the microstructure/nanostructure of the product cements is the key area to focus on when attempting to modulate drug release.

$$M_t = (2cDc_s t)^{1/2} \tag{6.2}$$

Equation (6.2) is the classical Higuchi equation, assuming pseudosteady-state conditions, where M_t is the cumulative drug release at time t, c is the initial drug concentration, c_s is the drug solubility and D is the diffusion coefficient.

$$M_t = A\left(\frac{2D_a c_s}{\tau\rho}t\right)^{1/2} \tag{6.3}$$

Equation (6.3) is a modified Higuchi equation to consider drug release from porous monolithic drug delivery systems,[46] where M_t is the cumulative drug release at time t, A is the drug concentration in the matrix, D_a is the drug diffusion coefficient in the aqueous phase, c_s is the drug solubility, τ is the tortuosity and ρ is the drug density.

In general, without considering the factors controlling drug release and the specific interactions between the drug molecules and the carrier system, as the carrier formulations are prepared with higher weight loadings of actives, the release rate is higher. Taking into account the Higuchi release kinetics, this can be expected since a higher initial drug loading results in a higher diffusion gradient between the carrier particles and the bulk medium, which, in turn, results in a higher coefficient of diffusion.

As the microstructure of the calcium phosphate cement products is key in determining active release rates, research has and should still focus on how the formation of the cement porous structure can be controlled and ideally designed to fit specific applications. One of the primary ways of achieving this control is by modifying the liquid:powder ratio during the cement-forming reaction. During an investigation into aspirin releasing cements,[41] it was found that there was a direct correlation between the amount of the dilute phosphoric used in the setting reaction and the aspirin release rate. More specifically, there was found to be a linear relationship between the product porosity and the liquid:powder ratio—as the liquid:powder ratio was increased, the aspirin release rate was also increased. This finding was significant as it provided for the first time a quantifiable means of modulating drug release rates by controlling the formation of the cement product. The more dilute the cement precursor particles are in suspension/solution, the larger are the voids that are formed between the developing crystalline aggregates.

The effect of modifying the liquid:powder ratio on release rates has been investigated using vancomycin,[47,48] ciprofloxacin,[48] 6-mercaptopurine[44] and bovine serum albumin,[28] amongst others. There is an important aspect to consider when changing the liquid:powder ratio as it could have implications for the application of these materials in the clinic. In terms of the product microstructure, a high liquid:powder ratio results in greater porosity—pore volumes, diameters, entrances and connectivity are all affected. As a result, release rates are increased, as is the total drug release as a fraction of the loaded amount. However, the increased cement volume due to the increased amount of the liquid phase will lead to longer periods of time for complete setting of the cement to take place. In addition, the increased porosity is certainly going to have an effect on the mechanical strength of the cement products. These extra details must be considered when designing a drug releasing cement system (Figures 6.6 and 6.7).

The impact the presence of the drug is going to have on the cement-forming process is another important feature to consider in the design of these drug delivery systems.

The pH of the system is an important parameter in determining the calcium phosphate phases formed. This in itself is a useful handle in modifying the drug release kinetics since porosity and thus drug release kinetics decrease in the order: HA > monetite > brushite.[49] pH 4.2 is a boundary for the formation of different calcium phosphate cement phases. At pH >4.2 HA is the thermodynamically favoured product, whilst at pH <4.2 brushite is the favoured calcium phosphate

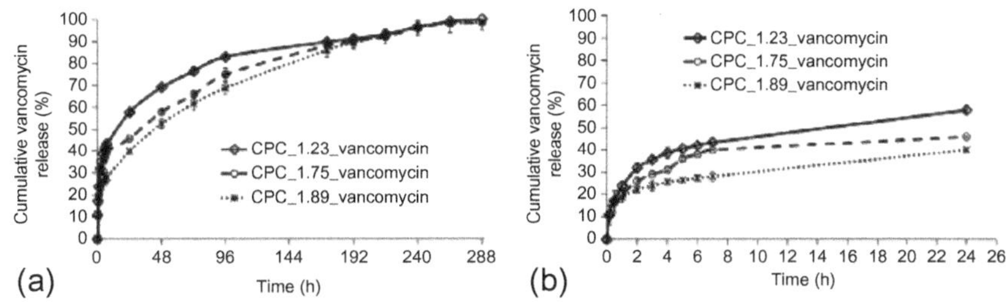

Figure 6.6 Vancomycin dissolution rates showing that an increased powder:liquid ratio (decreased liquid:powder ratio) results in a slower release rate. *(Reprinted from Ref. 47 with permission from Elsevier).*

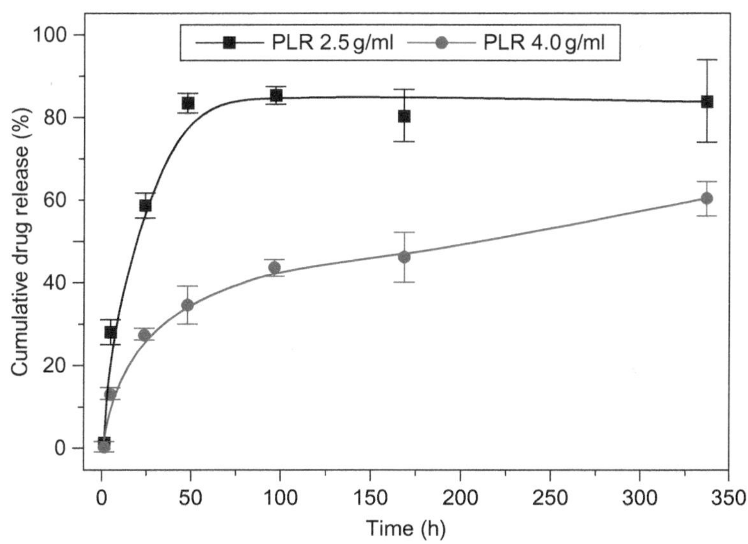

Figure 6.7 Dissolution study showing that as the liquid:powder ratio is increased, the rate of ciprofloxacin release increases. *(Reprinted from Ref. 48 with permission from Elsevier).*

phase. Therefore the inclusion of any drug compound to the cement-forming process which is going to alter the intended reaction pH significantly is going to have an impact not only on the products formed, but certainly on the drug release rate as well. Inclusion of ascorbic acid[50] and the carboxylic acids ciprofloxacin and cefaclor[51] was thought to alter the reaction pH significantly and thus affect the drug releasing properties of the cement products. However, it is possible that the presence of the carboxylic acid groups in ciprofloxacin and cefaclor enables these compounds to interact with the cement-forming products in a different way. By extension, purposeful manipulation of the reaction pH is of

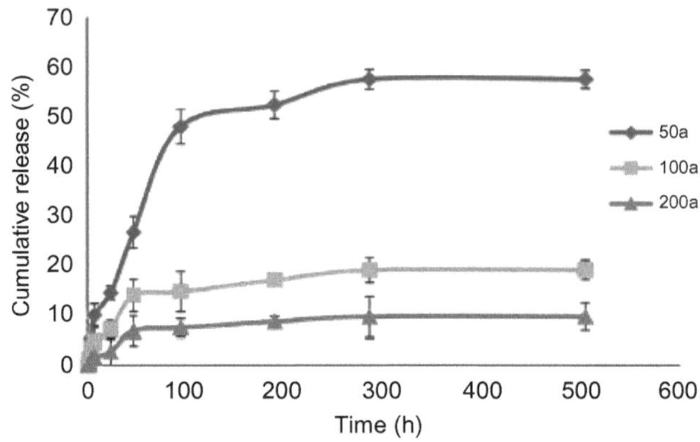

Figure 6.8 Release of varying amounts of ascorbic acid. The trend showing decreasing release rate with increasing drug concentration is counter-intuitive to a simple Noyes-Whitney type concentration-based diffusion gradient. *(Reprinted from Ref. 50 with permission from Elsevier).*

course an important method for controlling the drug release,[48] whilst an acidic environment within the cement leads to a greater chance of burst release and an increased rate of drug release thereafter (Figure 6.8).[52]

As was seen in section 6.2.1, control of cement precursor particle size is very important in determining product morphology—finer particles lead to smaller needle-like crystalline aggregates, whereas larger particles lead to larger plate-like crystalline aggregates. The needle crystals are packed more efficiently which results in a decreased inter-aggregate pore size. As a result, the tortuosity, pore diameter and pore volume of these products are decreased resulting in a decreased rate of active release.[48]

Furthermore, this effect was noted when drug release was studied from calcium phosphate cement systems with different macrostructures—planar surfaces as well as three-dimensional particulate systems.[44] Not completely applicable to the application of drug releasing calcium phosphate cement, but changing the material product geometry is a good method for controlling drug release kinetics from diffusion controlled, non-soluble drug delivery systems since the mean path length and material tortuosity can be changed dramatically.

6.2.5 Calcium Phosphate Cements—Additional Functionality

Despite being relatively new, the area of drug releasing calcium phosphate cements is currently a highly active area of research. Whilst methodologies are beginning to be formed on how to effectively control drug release rates, additional benefits of this approach to drug delivery are also being discovered which will improve the chances of these materials making it to market. Indomethacin (BCS class II) is an important active pharmaceutical and is often used as a model *in vitro* compound for the development of poorly water-soluble compound enabling formulations. It has been shown that the encapsulation of indomethacin within the porous structure of a calcium phosphate cement results in the dispersion of the drug in its amorphous form, as shown by X-ray diffraction and differential scanning calorimetry techniques. The formation of amorphous solid dispersions is highly dependent on the active being formulated, but the calcium phosphate cements could be an important addition to the formulation scientist's toolbox for delivering poorly water-soluble compounds.[53]

A common problem for the delivery of biomolecules such as proteins and cellular/viral vectors is that often the process of formulating them significantly affects the bioactivity of these types of active agents. In the context of treating bone defects, the proresorptive cytokine RANKL is an important molecule because it induces osteoclast activity. In combination with trehalose, a conservative agent, RANKL was adsorbed onto a brushite cement system and its biological activity retained up to 5 weeks after formulation.[54]

An under-researched, but potentially very significant area of drug releasing calcium phosphate cements, is the inclusion or substitution of other biologically important ions. Strontium is a particularly interesting ion to introduce, not only because its larger atomic radius (relative to calcium) leads to an increase in the porosity of cement products, but also because its *in situ* release could potentially have beneficial effects for the induction of hard tissue repair. Increasing mole% strontium substitution leads to an increase of the cement surface area and porosity, following which doxycycline hyclate was introduced to the strontium containing cement formulation by both *in situ* and post-impregnation methods. The formulation with doxycycline addition following cement formation demonstrated immediate drug release type kinetics, whereas the formulation produced with doxycycline present during the cement setting process had sustained release properties. The inclusion of strontium in conjunction with the appropriate doxycycline-loading method enables active release to occur up to 600 h. In consideration of the release kinetics and microstructural properties of these doxycycline releasing materials, it seems that the inclusion of strontium does not affect the drug releasing mechanism and kinetics directly, but rather affects the porosity of the cement products, as has been seen with variation of other cement-forming parameters (Figure 6.9).[55]

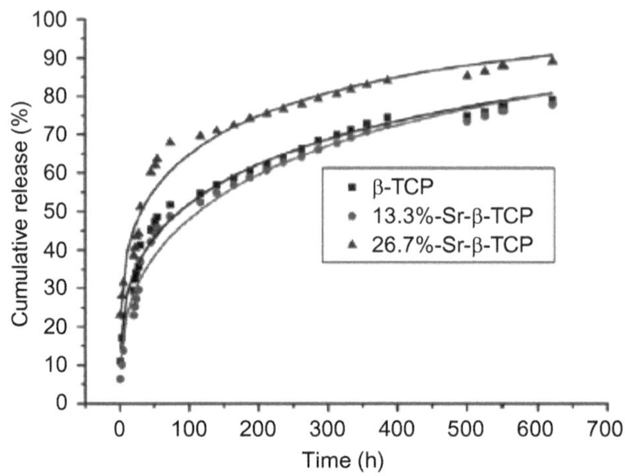

Figure 6.9 Increase in doxycycline release rate with increasing strontium substitution. *(Reprinted from Ref. 55 with permission from Elsevier).*

As the methods for producing and analyzing drug releasing calcium phosphate cements become more robust and reproducible, more *in vivo* studies on the efficacy of these new drug delivery systems need to be carried out. Only then can the potential of these innovative drug delivery systems be realized.

6.3 CALCIUM PHOSPHATE FOAMS FOR DRUG DELIVERY

As is the case with calcium phosphate cements, calcium phosphate foams are well known and studied for their potential use in creating osteoinductive, bioresorbable bone void fillers. Again, it has been only very recently that these systems are beginning to be investigated for their potential as drug delivery systems.

Calcium phosphate foams are characterized by their high porosity resulting from large macropores which have a high degree of interconnectivity. Many calcium phosphate foams have a hierarchical pore structure in that they contain not only macropores (>100 μm) and micropores (<100 nm), but also in some cases a set of pores with diameter approximately 1-10 μm. The high extent of macroporosity is particularly important in an *in vivo* setting as this provides a more favourable environment for encouraging cell colonization, vascularization and quick bone ingrowth, particularly into the very central parts of the filled defect/void. Human growth hormone (hGF)[56], hepatocyte growth hormone (HGF)[57] and recombinant human bone morphogenetic hormone-2 (rhBMP-2)[58] have all been shown to accelerate the rate of new bone growth when released from calcium phosphate scaffolds in comparison to the non active-loaded scaffolds. Though the presence and extent of the macroporosity is crucial to defining these foam systems, the presence of microporosity is also hugely important for encouraging new bone growth.[59] Most conventional foam fabrication methods such as the use of sacrificial porogens, solvent casting, gas/vacuum foaming, or heat-induced phase separation require some form of thermal treatment following the initial macrostructure forming step to: (i) consolidate the calcium phosphate foam crystalline backbone and/or (ii) remove any sacrificial pore-forming agent. In many cases this sintering step changes the morphology and phase purity of the foam structure and necessarily precludes drug incorporation during the foaming process (Figure 6.10).[2]

More recent methodology, however, seems to have found excellent solutions to these processing issues.[60] Macroporous self-setting calcium phosphate foams can be prepared by first suspending or dissolving a low-molecular weight surfactant into a typical cement-forming liquid phase and then adding this foamed liquid phase to a reactive calcium phosphate powder. The stability and structure of the foam are strongly dependent on the foaming agent.[61–63]

The inclusion of increasing amounts of doxycycline hyclate to these novel foam systems decreases the extent of porosity and the percent drug release, though is still higher than the equivalent cement. The rate of doxycycline release decreases proportionally with increasing drug-loading levels. This decrease in the release rate is a result of the decreased macroporosity, number of interconnected pores and pore entrances, as shown by porosimetry analysis (Figure 6.11).

6.4 DRUG RELEASE FROM CALCIUM PHOSPHATE COATINGS

Metallic alloys, and in particular those based on titanium are quickly becoming the gold standard substrate material for joint replacement and other conditions where the inclusion of load-bearing substrates is required (pins, screws, etc.). Whilst titanium alloys are known to be biocompatible, they are not bioactive, that is, they do not osseointegrate and form new chemical bonds between the implant surface and existing tissues *in situ*. Some form of surface engineering is therefore required to improve the properties of

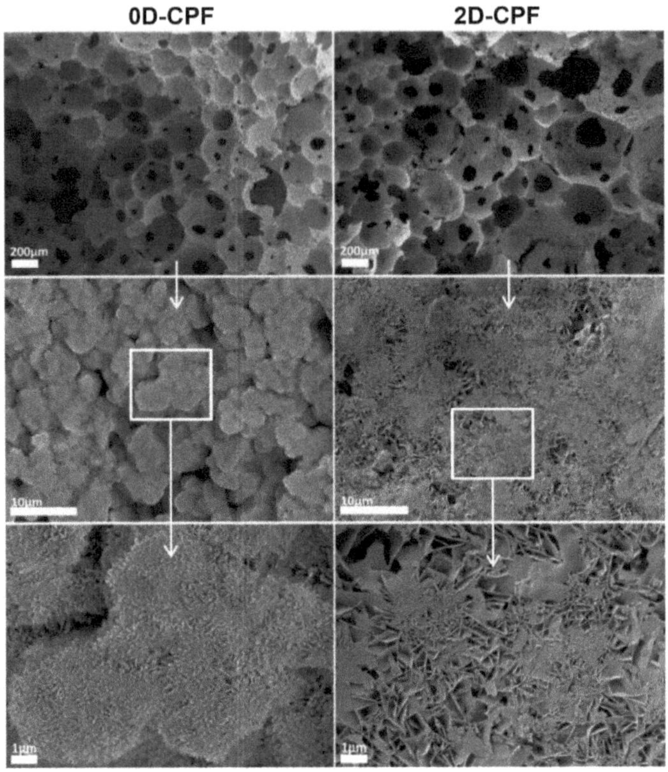

Figure 6.10 Hierarchical calcium phosphate foam porosity in comparison with a cement product (0D-CPF: 0 wt% doxycycline hyclate; 2D-CPF: 1.78 wt% doxycycline hyclate). *(Reprinted from Ref. 60 with permission from Elsevier).*

the titanium alloy prior to use. The use of calcium phosphate coatings in this respect can be advantageous in promoting long-lasting, durable fixation of the implant. One of the more serious consequences of this type of surgery is post-operative infection. One way of enhancing the efficiency of treating these types of infection is to include antibiotics into the calcium phosphate coatings applied to implant materials. In fact, the inclusion of antibiotics, along with growth factors such as bone morphogenetic proteins, is by far the most active area of research in the application of drug delivering calcium phosphate coatings.

Unfortunately, the development of processes for coating substrates with calcium phosphate materials often includes high temperatures, either during the process or as a post-process (e.g. plasma spray deposition, hot isostatic pressing, magnetron sputtering) which precludes the inclusion of active

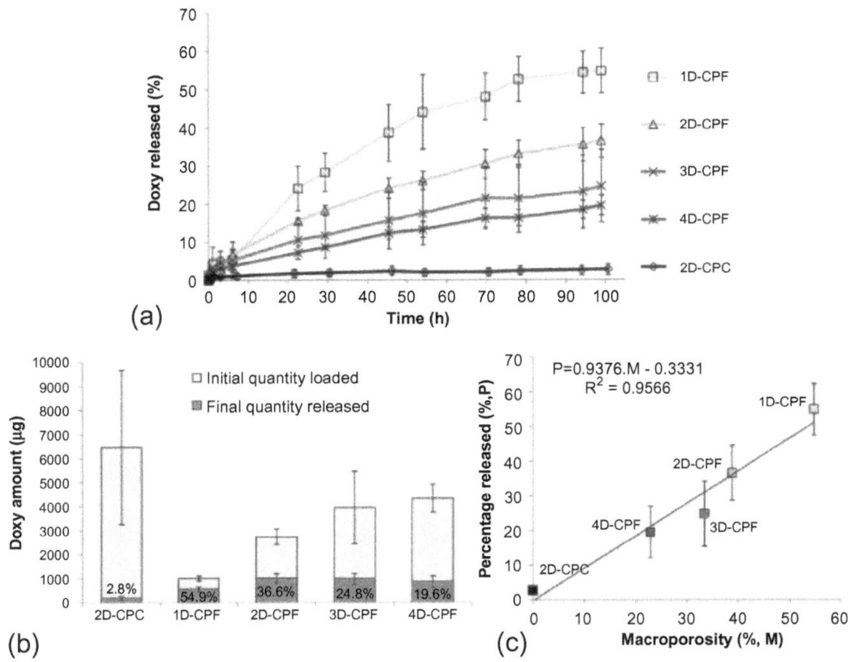

Figure 6.11 Correlation between the extent of macroporosity and doxycycline release. *(Reprinted from Ref. 60 with permission from Elsevier).*

drug molecules. Those coating techniques which are amenable to low-temperature processing such as sol-gel or dip-coating tend to have reduced interface bonding between the coating and the substrate and are prone to delamination.

As a result, two drug-loading techniques are prevalent in attempting to produce drug-loaded coatings. The first involves depositing the coating using conventional processing and then either 'dripping' concentrated active solutions onto the coated substrate or by immersing the substrate in a concentrated active solution for extended periods of time. Since by their nature, the coatings tend to be thin films, there is less reliance on diffusion-mediated processes to control active release and in fact the adsorption and desorption (release) processes are dominated by the specific interactions between the active and the calcium phosphate surface.[64] It can be seen therefore that surface properties, such as the specific surface area, are important in dictating the monolayer adsorption that can occur, whilst electrostatic interactions between the active and the calcium phosphate will determine the relative rates of adsorption and desorption from the loading and release media respectively. Calcium phosphate coatings are particularly beneficial in this respect, as compared with a silica surface for example, since the calcium phosphate phase contains both positive sites (Ca^{2+}), which can bind negatively charged or electronegative molecules, and negative sites (PO_4^-) which are suited to binding positively charged or electropositive molecules. From this it can therefore be seen that the exact ratio of calcium phosphate phases present in the coating can result in different distributions of charge along the coating surface and can thus have an important role in determining drug release.

Alternatively a method known as biomimetic co-precipitation is becoming increasingly popular.[65,66] This involves immersing the substrate in a supersaturated solution of simulated body fluid (SBF), in which typically 5-10 times the usual ion concentration is present. Biomimetic precipitation then occurs as apatite (typically CDHA) layers are precipitated onto the substrate surface. If the desired active is soluble in the supersaturated SBF solution, then as the apatite layer is precipitated, drug molecules are also co-precipitated (Figure 6.12).

Antibiotics containing carboxylic acid groups were found to have a higher rate of co-precipitation into the calcium phosphate coating, and then

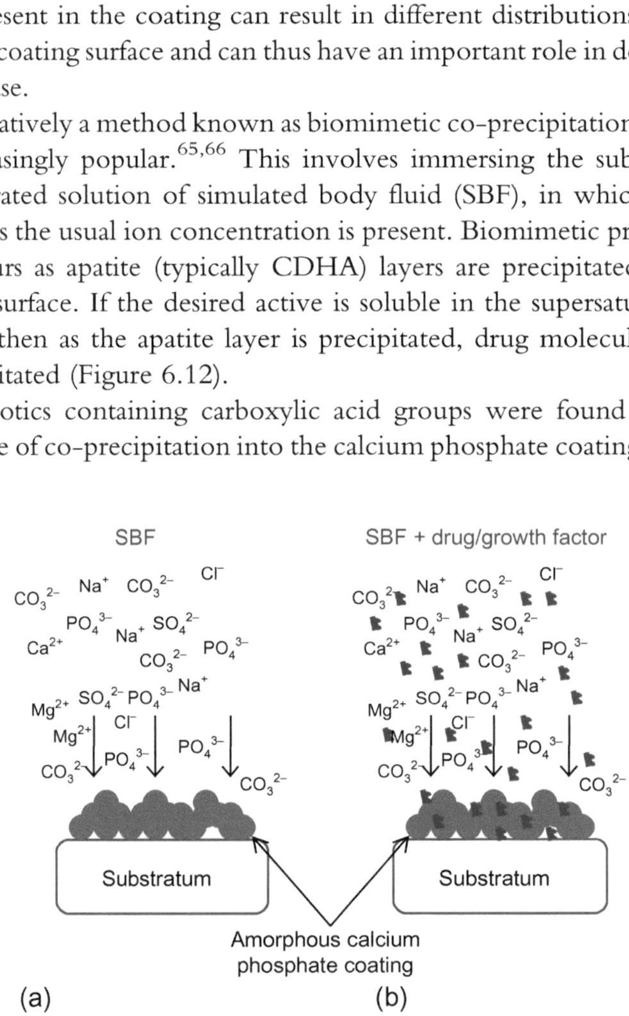

Figure 6.12 Schematic description of the biomimetic co-precipitation film-forming method: (a) no active present and (b) active present. *(Adapted from Ref. 2 with permission from Elsevier).*

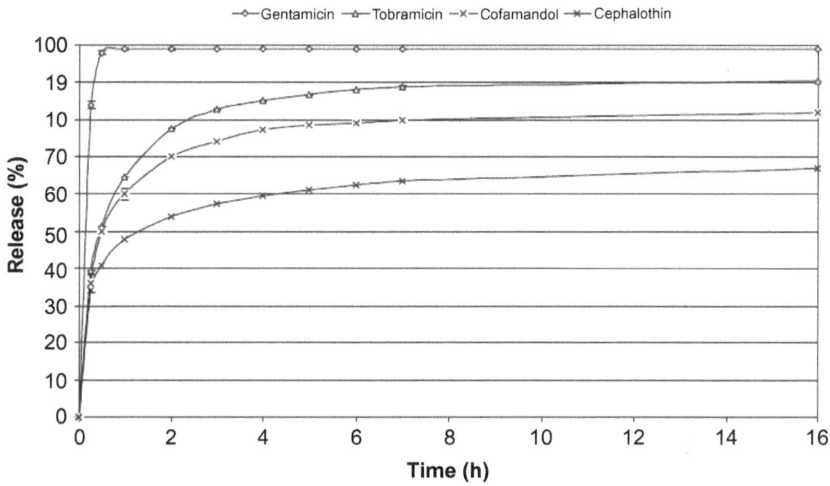

Figure 6.13 Dissolution profiles from a range of antibiotic-loaded calcium phosphate coatings. *(Reprinted from Ref. 67 with permission from Elsevier).*

also demonstrated slower release kinetics. It is thought that a chelating effect between the acidic groups and calcium ions in the coating is responsible for this favourable effect (Figure 6.13).[67]

There is no doubt that calcium phosphate-coated implants are an important class of biomaterials. A number of excellent studies have demonstrated the *in vitro* and *in vivo* efficacies of drug-loaded calcium phosphate coatings prepared by both adsorption and biomimetic co-precipitation routes, primarily for antibiotics[68–71] and growth factor[72–75] active delivery. Despite this, though, currently the only way of influencing the release rate is by altering the amount of drug present during the loading process. Until a robust method of controlling drug release through the formulation process is discovered, the use of calcium phosphate coatings as controlled release drug delivery systems may be limited.

REFERENCES

1. Forsén S, Kördel J. Calcium in biological systems. In: Bertini I, Gray HB, Lippard SJ, Valentine JS, editors. *Bioinorganic chemistry*. Mill Valley: University Science Books; 1994. p. 107–8.
2. Bose S, Tarafder S. Calcium phosphate ceramic systems in growth factor and drug delivery for bone tissue engineering: a review. *Acta Biomater* 2012;**8**(4):1401–21.
3. Carrodeguas R, De Aza S. α-Tricalcium phosphate: synthesis, properties and biomedical applications. *Acta Biomater* 2011;**7**(10):3536–46.

4. Dorozhkin SV, Epple M. Biological and medical significance of calcium phosphates. *Angew Chem Int Edit* 2002;**41**(17):3130–46.
5. Bose S, Tarafder S, Edgington J, Bandyopadhyay A. Calcium phosphate ceramics in drug delivery. *JOM-US* 2011;**63**(4):93–8.
6. Chow LC. Development of self-setting phosphate cements. *J Ceram Soc Jpn* 1991;**91** (10):954–64.
7. LeGeros RZ. Biodegradation and bioresorption of calcium phosphate ceramics. *Clin Mater* 1993;**14**(1):65–88.
8. Bohner M. Calcium orthophosphates in medicine: from ceramics to calcium phosphate cements. *Injury* 2000;**31**(Suppl. 4):37–47.
9. Vallet-Regi M, Gonzalez-Calbert JM. Calcium phosphates as substitution of bone tissues. *Prog Solid State Chem* 2004;**32**:1–31.
10. Dorozhkin SV. Amorphous calcium (ortho)phosphates. *Acta Biomater* 2010;**6** (12):4457–75.
11. Cicek G, Aksoy EA, Durucan C, Hasirci N. Alpha-tricalcium phosphate (α-TCP): solid state synthesis from different calcium precursors and the hydraulic reactivity. *J Mater Sci Mater Med* 2011;**22**(4):809–8117.
12. Dorozhkin SV. Self-setting calcium orthophosphate formulations: cements, concretes, pastes and putties. *Int J Mater Chem* 2012;**1**(1):1–48.
13. Dorozhkin SV. A detailed history of calcium orthophosphates from 1770s till 1950. *Mater Sci Eng C* 2013;**33**(6):3085–110.
14. Albee FH, Morrison HF. Studies in bone growth: triple calcium phosphate as a stimulus to osteogenesis. *Ann Surg* 1920;**71**(1):32–9.
15. McClendon JF, Carpousis A. Prevention of dental caries by brushing the teeth with powdered fluorapatite. *J Dent Res* 1945;**24**:199.
16. Kingery WM. Fundamental study of phosphate bonding in refractories: 2, cold-setting properties. *J Am Ceram Soc* 1950;**33**(8):242–6.
17. Köster K, Karbe K, Kramer H, Heide H, König R. Experimenteller Knochenersatz durch resorbierbare Calciumphosphat-Keramik. *Langenbeck Arch Chir* 1976;**341**(2):77–86.
18. Brown WE, Chow LC. *J Dent Res* 1983;**62**(Suppl. 1):672.
19. LeGeros RZ. *J Dent Res* 1982;**61**:343.
20. Gruninger SE, Siew C, Chow LC, O'Young A, Tsao NK, Brown WE. Evaluation of the biocompatibility of a new calcium phosphate setting cement. *J Dent Res* 1984;**63**(Spec. Iss.):200.
21. Baroud G, Swanson T, Steffen T. Setting properties of four acrylic and two calcium-phosphate cements used in vertebroplasty. *J Long Term Eff Med* 2006;**16**(1):51–9.
22. Fernández E, Ginebra M-P, Bermudez O, Boltong MG, Driessens FCM, Planell JA. Dimensional and thermal behaviour of calcium phosphate cements during setting compared to PMMA bone cements. *J Mater Sci Lett* 1995;**14**(1):4–5.
23. Ishikawa K. Calcium phosphate cement. In: Kokubu T, editor. *Bioceramics and their clinical applications*. Cambridge: Woodhead Publishing Ltd; 2008. p. 438–63.
24. Mirtchi AA, LeMaître J, Hunting E. Calcium phosphate cements: action of setting regulators on the properties of the β-tricalcium phosphate-monocalcium phosphate cements. *Biomaterials* 1989;**10**(9):634–8.
25. Mirtchi AA, LeMaître J, Hunting E. Calcium phosphate cements: study of the β-tricalcium phosphate—monocalcium phosphate system. *Biomaterials* 1989;**10** (7):475–80.
26. Ginebra M-P, Fernández A, Driessens FCM, Planell JA. Modeling of the hydrolysis of α-tricalcium phosphate. *J Am Ceram Soc* 1999;**82**(10):2808–12.
27. Ginebra M-P, Canal C, Espanol M, Pastorino D, Montufar EB. Calcium phosphate cements as drug delivery materials. *Adv Drug Deliv Rev* 2012;**64**(12):1090–110.

28. Espanol M, Perez R, Montufar EB, Marichal C, Sacco A, Ginebra M-P. Intrinsic porosity of calcium phosphate cements and its significance for drug delivery and tissue engineering applications. *Acta Biomater* 2009;**5**(7):2752–62.
29. Zhang J, Tancret F, Bouler J. Fabrication and mechanical properties of calcium phosphate cements (CPC) for bone substitution. *Mater Sci Eng C* 2011;**31**(4):740–7.
30. Rajzer I, Castaño O, Engel E, Planell JA. Injectable and fast resorbable calcium phosphate cement for body-setting bone grafts. *J Mater Sci Mater Med* 2010;**21**(7):2049–56.
31. Takagi S, Chow LC, Hirayama S, Sugawara A. Premixed calcium-phosphate cement pastes. *J Biomed Mater Res B* 2003;**67**(2):689–96.
32. Carey LE, Xu HHK, Simon Jr. CG, Takagi S, Chow LC. Premixed rapid-setting calcium phosphate composites for bone repair. *Biomaterials* 2005;**26**(24):5002–14.
33. Vorndran E, Geffers M, Ewald A, Lemm M, Nies B, Gbureck U. Ready-to-use injectable calcium phosphate bone cement paste as drug carrier. *Acta Biomater* 2013;**9**(12):9558–67.
34. Heinemann S, Rössler S, Lemm M, Ruhnow M, Nies B. Properties of injectable ready-to-use calcium phosphate cement based on water-immiscible liquid. *Acta Biomater* 2013;**9**(4):6199–207.
35. Bohner M, van Lenthe GH, Grünenfelder S, Hirsiger W, Evison R, Müller R. Synthesis and characterization of porous beta-tricalcium phosphate blocks. *Biomaterials* 2005;**26**(31):6099–105.
36. Brown WE, Chow LC. Combinations of sparingly soluble calcium phosphates in slurries and pastes as mineralizers and cements. US patent 4612053 A; 1986.
37. Otsuka M, Matsuda Y, Yu D, Wong J, Fox JL, Higuch WI. A novel skeletal drug delivery system for anti-bacterial drugs using self-setting hydroxyapatite cement. *Chem Pharm Bull* 1990;**38**(12):3500–2.
38. Wong J, Otsuka M, Higuchi WI, Powell GL, Fox JL. Effect of laser irradiation on the dissolution kinetics of hydroxyapatite preparations. *J Pharm Sci* 1990;**79**(6):510–5.
39. Otsuka M, Matsuda Y, Kokubo T, Yoshihari S, Nakamura T, Yamamura T. A novel skeletal drug delivery system using self-setting bioactive glass bone cement IV: cephalexin release from cement containing polymer-coated bulk powder. *Biomed Mater Eng* 1993;**3**(4):229–36.
40. Otsuka M, Nakahigashi Y, Matsuda Y, Fox JL, Higuchi WI, Sugiyami Y. Effect of geometrical cement size on in vitro and in vivo indomethacin release from self-setting apatite cement. *J Control Release* 1998;**52**(3):281–9.
41. Otsuka M, Matsuda Y, Suwa Y, Fox JL, Higuchi WI. A novel skeletal drug-delivery system using self-setting calcium phosphate cement. 4. Effects of the mixing solution volume on the drug-release rate of heterogeneous aspirin-loaded cement. *J Pharm Sci* 1994;**83**(2):259–63.
42. Otsuka M, Matsuda Y, Suwa Y, Fox JL, Higuchi WI. A novel skeletal drug-delivery system using self-setting calcium phosphate cement. 3. Physicochemical properties and drug-release rate of bovine insulin and bovine albumin. *J Pharm Sci* 1994;**83**(2):255–8.
43. Otsuka M, Matsuda Y, Suwa Y, Fox JL, Higuchi WI. A novel skeletal drug delivery system using a self-setting calcium phosphate cement. 5. Drug release behavior from a heterogeneous drug-loaded cement containing an anticancer drug. *J Pharm Sci* 1994;**83**(11):1565–8.
44. Otsuka M, Matsuda Y, Fox JL, Higuchi WI. A novel skeletal drug delivery system using self-setting calcium phosphate cement. 9: effects of the mixing solution volume on anticancer drug release from homogeneous drug-loaded cement. *J Pharm Sci* 1995;**84**(6):733–6.

45. Gbureck U, Vorndran E, Barralet J. Modeling vancomycin release kinetics from microporous calcium phosphate ceramics comparing static and dynamic immersion conditions. *Acta Biomater* 2008;**4**(5):1480–6.
46. Wise DL. *Handbook of pharmaceutical controlled release technology*. New York: Marcel Dekker Inc.; 2000
47. Loca D, Sokolova M, Locs J, Smirnova A, Irbe Z. Calcium phosphate bone cements for local vancomycin delivery. *Mater Sci Eng C* 2015;**49**:106–13.
48. Hofmann MP, Mohammed AR, Perrie Y, Gbureck U, Barralet JE. High-strength resorbable brushite bone cement with controlled drug-releasing capabilities. *Acta Biomater* 2009;**5**(1):43–9.
49. Gbureck U, Vorndran E, Müller FA, Barrelet JE. Low temperature direct 3D printed bioceramics and biocomposites as drug release matrices. *J Control Release* 2007;**122**(2):173–80.
50. Hemmati K, Hesaraki D, Nemati A. Evaluation of ascorbic acid-loaded calcium phosphate bone cements: physical properties and in vitro release behaviour. *Ceram Int* 2014;**40**(3):3961–8.
51. Akashi A, Matsuya Y, Unemori M, Akamine A. Release profile of antimicrobial agents from α-tricalcium phosphate cement. *Biomaterials* 2001;**22**:2713–7.
52. Irbe Z, Loca D, Vempere D, Berzina-Cimdina L. Controlled release of local anesthetic from calcium phosphate bone cements. *Mater Sci Eng C* 2012;**32**(6):1690–4.
53. Otsuka M, Matsuda Y, Suwa Y, Fox JL, Higuchi WI. A novel skeletal drug delivery system using self-setting calcium phosphate cement. 2. Physicochemical properties and drug release rate of the cement-containing indomethacin. *J Pharm Sci* 1994;**83**(5):611–5.
54. Le Nihouannen D, Komarova SV, Gbureck U, Barralet JE. Bioactivity of bone resorptive factor loaded on osteoconductive matrices: stability post-dehydration. *Eur J Pharm Biopharm* 2008;**70**(3):813–8.
55. Alkhraiset MH, Rueda C, Cabrejos-Azama J, Lucas-Aparicio J, Mariño FT, Torres García-Denche J, et al. Loading and release of doxycycline hyclate from strontium-substituted calcium phosphate cement. *Acta Biomater* 2010;**6**(4):1522–8.
56. Guicheux J, Gauthier O, Aguado E, Pilet P, Couillaud S, Jegou D, et al. Human growth hormone locally released in bone sites by calcium-phosphate biomaterial stimulates ceramic bone substitution without systemic effects: a rabbit study. *J Bone Miner Res* 1998;**13**(4):739–48.
57. Hossain M, Irwin R, Baumann MJ, McCabe LR. Hepatocyte growth factor (HGF) adsorption kinetics and enhancement of osteoblast differentiation on hydroxyapatite surfaces. *Biomaterials* 2005;**26**(15):2595–602.
58. Koempel JA, Patt BS, O'Grady K, Wozney J, Toriumi DM. The effect of recombinant human bone morphogenetic protein-2 on the integration of porous hydroxyapatite implants with bone. *J Biomed Mater Res* 1998;**41**(3):359–63.
59. Woodard JR, Hilldore AJ, Lan SK, Park CJ, Morgan AW, Eurell JAC, et al. The mechanical properties and osteoconductivity of hydroxyapatite bone scaffolds with multi-scale porosity. *Biomaterials* 2007;**28**(1):45–54.
60. Pastorino D, Canal C, Ginebra M-P. Drug delivery from injectable calcium phosphate foams by tailoring the macroporosity-drug interaction. *Acta Biomater* 2015;**12**:250–9.
61. Montufar EB, Traykova T, Gil C, Harr I, Almirall A, Aguirre A, et al. Foamed surfactant solution as a template for self-setting injectable hydroxyapatite scaffolds for bone regeneration. *Acta Biomater* 2010;**6**:876–85.
62. Montufar EB, Traykova T, Planell JA, Ginebra M-P. Comparison of a low molecular weight and a macromolecular surfactant as foaming agents for injectable self setting hydroxyapatite foams: polysorbate 80 versus gelatine. *Mater Sci Eng C* 2011;**31**(7):1498–504.

63. Ginebra M-P, Delgado J-A, Harr I, Almirall A, Del Valle S, Planell JA. Factors affecting the structure and properties of an injectable self-setting calcium phosphate foam. *J Biomed Mater Res A* 2007;**80A**(2):351–61.
64. Palazzo B, Iafisco M, Laforgia M, Margiotta N, Natile G, Bianchi CL, et al. Biomimetic hydroxyapatite—drug nanocrystals as potential bone substitutes with antitumor drug delivery properties. *Adv Funct Mater* 2007;**17**(13):2180–8.
65. Barrere F, van Blitterswijk CA, de Groot K, Layrolle P. Influence of ionic strength and carbonate on the Ca-P coating formation from SBF × 5 solution. *Biomaterials* 2002;**23**(9):1921–30.
66. Habibovic P, Barrère F, van Blitterswijk CA, de Groot K, Layrolle P. Biomimetic hydroxyapatite coating on metal implants. *J Am Ceram Soc* 2002;**85**(3):517–22.
67. Stigter M, Bezemer J, de Groot K, Layrolle P. Incorporation of different antibiotics into carbonated hydroxyapatite coatings on titanium implants, release and antibiotic efficacy. *J Control Release* 2004;**99**(1):127–37.
68. Roy M, Bandyopadhyay A, Bose S. In vitro antimicrobial and biological properties of laser assisted tricalcium phosphate coating on titanium for load bearing implant. *Mater Sci Eng C* 2009;**29**(6):1965–8.
69. Elayaraja K, Chandra VS, Joshy MIA, Suganthi RV, Asokan K, Kalkura SN. Nanocrystalline biphasic resorbable calcium phosphate (HAp/β-TCP) thin film prepared by electron beam evaporation technique. *Appl Surf Sci* 2013;**274**:203–9.
70. Kazemzadeh-Narbat M, Kindrachuk J, Duan K, Jenssen H, Hancock REW, Wang R. Antimicrobial peptides on calcium phosphate-coated titanium for the prevention of implant-associated infections. *Biomaterials* 2010;**31**(36):9519–26.
71. Radin S, Campbellt JT, Ducheyne P, Cuckler JM. Calcium phosphate ceramic coatings as carriers of vancomycin. *Biomaterials* 1997;**16**(11):777–82.
72. Liu Y, de Groot K, Hunziker EB. BMP-2 liberated from biomimetic implant coatings induces and sustains direct ossification in an ectopic rat model. *Bone* 2015;**36**(5):745–57.
73. Tsurushima H, Marushima A, Suzuki K, Oyane A, Sogo Y, Nakamura K, et al. Enhanced bone formation using hydroxyapatite ceramic coated with fibroblast growth factor-2. *Acta Biomater* 2010;**6**(7):2751–9.
74. Ito Y, Tsurushima H, Sato M, Ito A, Oyane A, Sogo Y, et al. Angiogenesis therapy for brain infarction using a slow-releasing drug delivery system for fibroblast growth factor 2. *Biochem Biophys Res Commun* 2013;**432**(1):182–7.
75. Majid K, Tseng MD, Baker KC, Reyes-Trocchia A, Herkowitz HN. Biomimetic calcium phosphate coatings as bone morphogenetic protein delivery systems in spinal fusion. *Spine J* 2011;**11**(6):560–7.

CHAPTER 7

Alternative Inorganic Systems for Controlled Release Applications

Contents

7.1 Introduction 189
7.2 Clays 190
  7.2.1 Anionic clays: Layered double hydroxides 190
    7.2.1.1 Synthetic considerations and active loading 192
    7.2.1.2 Incorporation of non-ionic compounds 194
    7.2.1.3 Controlling active release from LDHs 198
  7.2.2 Cationic clays 201
    7.2.2.1 Cationic clays as inorganic CRT systems 203
    7.2.2.2 Smectite clays for inorganic CRT 204
    7.2.2.3 Other clays for inorganic CRT 206
7.3 Alternative Mesoporous Inorganic CRT Materials 207
  7.3.1 Mesoporous zirconia 207
  7.3.2 Mesoporous alumina 208
7.4 Nanoparticle Delivery Systems 211
  7.4.1 Iron oxide nanoparticles 211
  7.4.2 Gold nanoparticles 214
References 216

7.1 INTRODUCTION

Up to this point we have only considered the materials chemistry of silicate-based glasses (Chapters 4 and 5) and calcium phosphate-based ceramics (Chapter 6) for inorganic controlled release technology (inorganic CRT) applications. These two classes of materials are the main focus of the research being carried out into inorganic CRT formulations. Within these two groups of formulations the majority of release systems developed are based on non-soluble porous matrix type systems. As the benefits of inorganic systems for controlled release applications are beginning to be more widely recognized, the use of alternative delivery systems is gaining more traction amongst pharmaceutical formulation and materials science researchers.

Inorganic Controlled Release Technology
http://dx.doi.org/10.1016/B978-0-08-099991-3.00007-7

Considering the spectrum of materials that could potentially be used for controlled release applications, it is slightly surprising that silicate- and phosphate-based materials have dominated the field to date. Research into the use of other inorganic systems for controlled release formulations is ongoing however and novel systems with specific advantages over traditional polymer-based controlled delivery systems are being developed. Not only are new chemistries being developed for these systems, but new release systems are also being developed which rely on alternative release mechanisms, expanding on the range of potential therapeutic applications inorganic CRT systems could be used for. This chapter will highlight some of the more well-studied and alternative inorganic CRT systems.

7.2 CLAYS

Clays are a group of naturally formed silicate-based minerals which typically also consist of aluminium.[1,2] There are exceptions to this however and most clay minerals are composed of varying amounts of other cationic species, in particular iron, magnesium and group 1 and 2 metals. Talc, a specific type of clay mineral, is composed entirely of hydrated magnesium silicate.

Clays are known as phyllosilicate minerals which are a reference to their layered structure (see Figure 7.1), and it is the physical and chemical nature of this layered structure, or more specifically, the *interlayer space* which is instrumental in determining the ion exchange potential of the material. Clay's ion exchange potential amongst other favourable physical properties has meant that these minerals have been routinely used as formulation excipients to perform a number of different functions: diluent, emulsifier, anticaking and even as active agent carriers.[3] This active modifying property was only empirically observed at first,[4,5] but efforts have now been focused on evaluating the specific properties of different subgroups of clay minerals and designing controlled release systems based on specific active-clay interactions.

7.2.1 Anionic Clays: Layered Double Hydroxides

Amongst the huge number of natural clay materials, anionic clays, those that have overall positively charged layers and interstitially situated anions and water, are rarer in nature than their cationic counterparts; though their synthesis is relatively straightforward (see Section 7.2.2). Depending upon the exact nature of their composition, these clay materials are more commonly referred to as hydrotalcites (a specific anionic clay with composition $Mg_6Al_2CO_3(OH)_{16}{\cdot}4(H_2O)$), hydrotalcite-like, or more generally, layered double hydroxides (LDHs).[6]

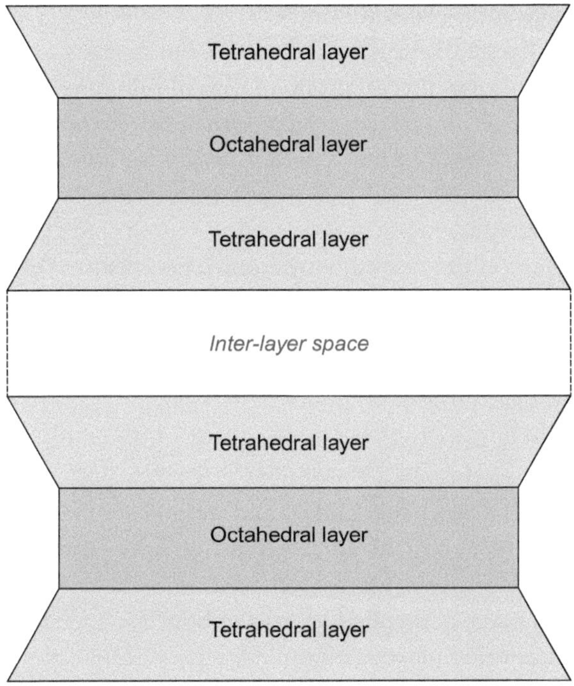

Figure 7.1 Schematic of a typical 2:1 natural clay (see Section 7.3 for more detail on this nomenclature).

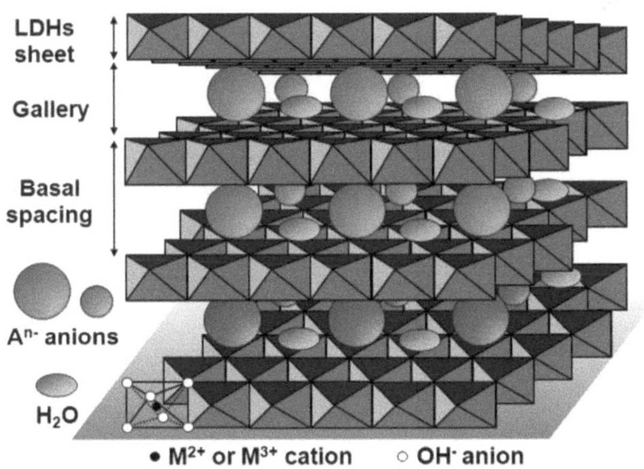

Figure 7.2 General structure of a layered double hydroxide. *(Reprinted from Ref. 12 under a creative commons attribution licence, https://creativecommons.org/licenses/by/3.0).*

As the name suggests, the structure forming part of LDHs are sheets composed of mixed hydroxides, generally based on $Mg(OH)_2$ (brucite) (see Figure 7.2). In brucite minerals the positive charge derived from the Mg^{2+} ion is satisfied by octahedrally coordinated hydroxide groups.

Isomorphic substitution (substitution of an element which does not give rise to a significant change in the crystal structure) of magnesium for a trivalent cation such as Al^{3+} gives the structure an overall imbalance in charge which is then quenched by the intercalation of anions between the magnesium-based layers. In naturally derived LDHs typical intercalating anions include carbonate (CO_3^{2-}) and hydroxide (OH^-), but synthetically it is possible to introduce any number of different anionic (and other) species. It is this property of being able to host molecules between the structural layers that makes LDHs such an attractive material for inorganic controlled release applications.[7–11]

7.2.1.1 Synthetic Considerations and Active Loading

There are a number of structural variables that can be altered to influence the properties of LDHs, see Table 7.1, though very few of these alterations are found in naturally formed LDH products.[13] All however have been applied to the preparation of synthetic LDHs, and many have been investigated in the context of influencing the release rates of intercalated actives, which gives the materials scientist/controlled release formulator a huge array of flexibility in designing controlled release systems for specific applications.

There are three principle mechanisms by which LDH materials can be prepared and/or modified to prepare alternative LDH products. All three of these processing routes can be modified to include active molecules and each approach has its own merits and benefits for producing LDH inorganic CRT systems.

Anionic exchange is the exchange of the layer stabilizing anions already resident within the LDH with alternative anions in particular, though methods are reported for exchanging in neutral molecules. In a practical sense, it is as simple a process as stirring the LDH powder in a concentrated solution of the desired substituting species, as shown in Figure 7.3. However, on the molecular level, achieving effective exchange is significantly more difficult and is highly dependent on the variables listed in Table 7.1, in particular the size

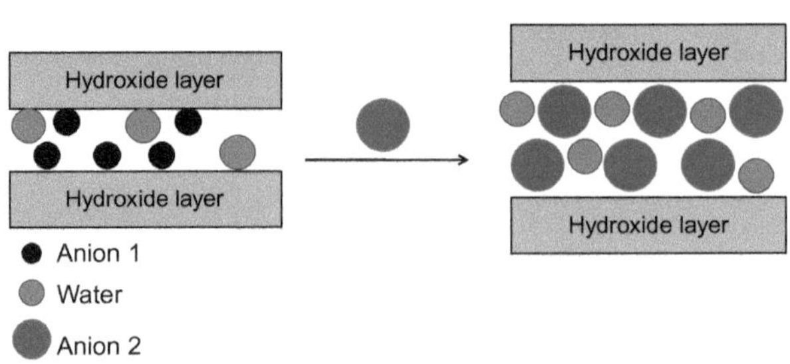

Figure 7.3 Schematic representation of an ion exchange process.

Table 7.1 LDH synthetic variables giving rise to control over active release

Structural variable	Preparation variable
Cation size	pH
Ratio of cations	Precipitation method
Cation stereochemistry	Reagent concentration
Cation mixture	Temperature and ageing
Nature of the anion	Washing and drying
	Presence of impurities

Adapted from Refs. 13,14.

and charge density of the incoming anion and the concentration and pH of the exchanging medium. It is reported that LDH-type materials prefer anions with higher charge densities to be intercalated,[15,16] so as a result LDH materials containing nitrates and halides are preferred exchange precursors since their exchange potential is relatively high.

It was by an anionic exchange process that these materials were first loaded with actives for biomedical purposes.[17–19] Consideration must be given to the specific active required to be intercalated. Since charge stabilizing anionic species are being lost from the structures, the incoming active must be negatively charged or have significant hydrogen bond donating character, common examples are polyhydroxyl or polyamino compounds. Carboxylic acids and amino acids are suitable for anionic exchange since both types of compounds are ionized at pH levels above ~3.5-4.5. By extension, most biological molecules such as peptides, nucleotides and even strands of DNA are also effective for intercalation in LDHs since they are highly negatively charged in water at all but very low pH levels.

Coprecipitation is a means of both preparing bespoke synthetic LDHs but also a way of introducing the active of choice during the carrier synthesis. Typically, nitrates of the desired product hydroxides (Mg, Al and Zn are the most common) are dissolved in aqueous solution with the active and the pH slowly raised by the addition of concentrated alkali hydroxides or through a controlled degradation of urea. Once the pH reaches above ~9, the LDH phases begin to precipitate out of solution. As already mentioned, nitrates are the most common precursor anion species as they exchange most readily. This is a relatively simple method of varying the hydroxide content of the LDH products, simply by varying the molar ratio of the species in the precursor solution. There are limitations here though, the active must be water soluble (as with all simple LDH-active

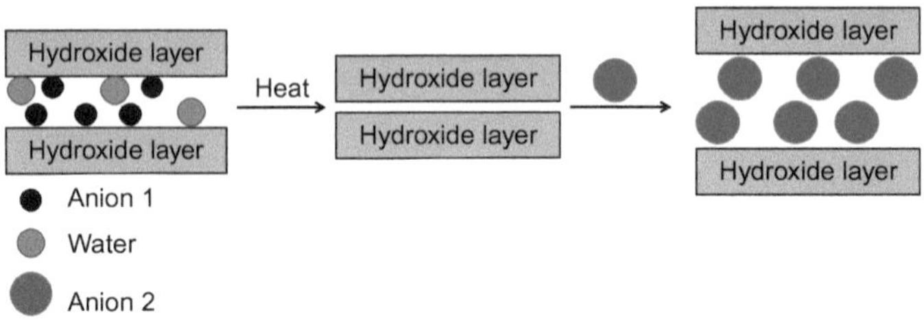

Figure 7.4 Schematic representation of the preparation of drug-LDH composites by the reconstruction process.

conjugates) and it must be stable to hydrolytic decomposition in high pH environments.

Reconstruction is a process based on a particularly beneficial property which LDHs possess – the memory effect.[20,21] Mild calcination of an LDH containing intercalated anions at 500–600 °C under an inert gas atmosphere results in an oxide product with a charge imbalance. As such, upon rehydration the calcined oxide 'remembers' its previous physical condition and returns to its LDH conformation. The presence of the desired species to be intercalated within the rehydrating solution produces the drug-loaded LDH (see Figure 7.4). It is reported that active intercalation by the reconstruction process is a much more facile process than anionic exchange, presumably because the thermodynamic preference for stabilizing the residual positive charges in the calcined oxide structure is far greater than that of replacing existing intercalating ions.

7.2.1.2 Incorporation of Non-Ionic Compounds

A major limitation of LDH inorganic CRT systems as described so far is that it is only possible to incorporate negatively charged or highly polar compounds within the lamellar LDH structure. This is obviously very limiting as these types of compounds are by definition water soluble. However, methodology does exist for the intercalation of neutral compounds regardless of their solubility property. This additional flexibility in the types of actives that can be added makes LDHs very versatile drug delivery systems.

The use of LDHs to intercalate and improve the dissolution properties of the poorly water-soluble, negatively charged compounds ketoprofen and indomethacin has already been described.[22] Rapid dissolution and degradation of the LDH matrix at acidic pH ($pH < 4$) resulted in drug supersaturation and improved the dissolution profiles in acidic media *in vitro*, the assumption being that improved bioavailability *in vivo* would also occur. Building on this, Tyner

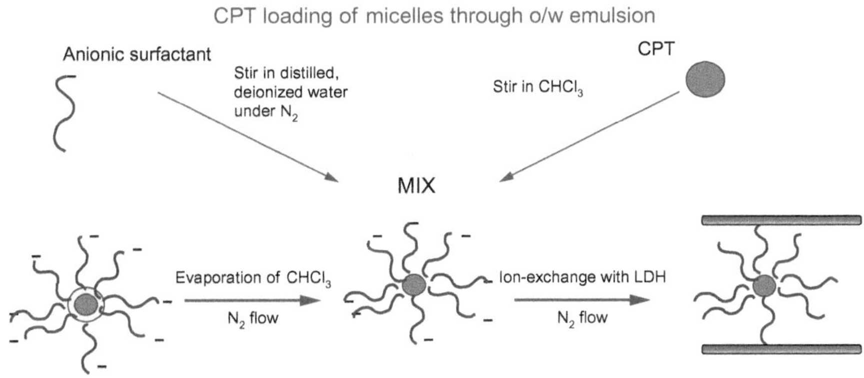

Figure 7.5 Preparation of LDH-camptothecin drug delivery system. *(Reprinted from Ref. 23 with permission from Elsevier).*

first described an approach involving the incorporation of camptothecin (CPT) within the biocompatible anionic surfactants sucrose aspartate and sodium cholate.[23] This negatively charged drug-surfactant complex was then intercalated into an LDH matrix by a standard ion exchange. The overall process is represented schematically in Figure 7.5.

Successful intercalation of the drug-loaded micelle was proven by powder X-ray diffraction (XRD) analysis, which showed a basal layer spacing of 8.8 Å, for the LDH-NO_3 powder and interlayer distances of 32.7 and 29.8 Å for the LDH-drug-loaded sodium cholate nanohybrid and the LDH-drug-loaded sucrose aspartate nanohybrid complex, respectively. Complete camptothecin release was noted within 10 min from the LDH-sucrose aspartate delivery system at both pH 4.8 and 7.2, which represents a three-fold increase in *in vitro* solubilization as compared with a similar analysis carried out on camptothecin only (see Figure 7.6). As an additional formulation benefit, the authors suggest

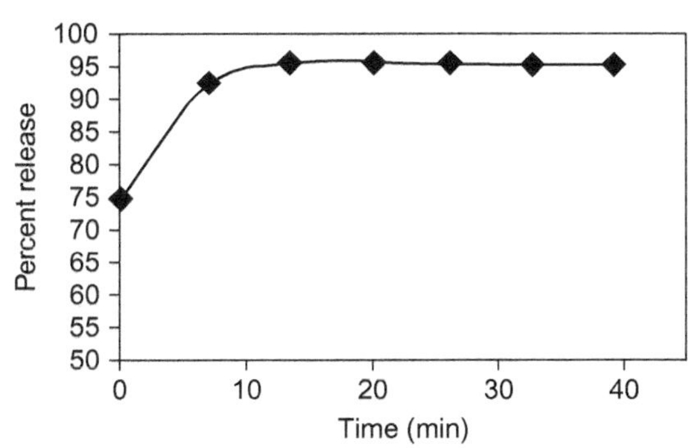

Figure 7.6 Camptothecin release from LDH-micelle complex in pH 4.8 buffered release media. *(Reprinted from Ref. 21 with permission from Elsevier).*

that as the overall LDH-drug-micelle complex is charge-neutral the nanohybrid complexes should be effectively endocytosed by cells.

In a similar approach, Kameshima investigated the absorption of sodium oleate into the interlayer space of LDHs and also the adsorption of the oleate molecules onto the surface of the LDH particles.[24] This was one of the earliest studies into how variation of the LDH forming process affects the product structure and as a consequence provides a handle on varying active release. Kameshima found that as the Mg/Al ratio decreased, the specific surface area and the pore volume (both measured by BET gas adsorption, see Chapter 3) of the LDH products generally increased. In addition, the reduction in the Mg/Al ratio resulted in an increased capacity of the LDH products to intercalate oleate molecules. This effect was seen for LDH products prepared by both coprecipitation and reconstruction reaction processes. Again, a specific increase in the basal layer spacing as determined by XRD suggested that the oleate molecules were either intercalated in the anionic layers as bilayers or as micelle structures. As was mentioned previously, intercalation of ions into reconstructed LDHs is a much more efficient process, and this was the case here, very little oleate intercalation was seen by the ion exchange method. Interestingly oleate ions were also found to be adsorbed to the LDH particle surfaces, which has important implications for the formulation of LDH-drug composite delivery systems overall. Oleate is resistant to solubilization under acidic conditions meaning that if the LDH surfaces are significantly protected by oleate ions then these systems may pass through the acidic environment of the stomach relatively unharmed. Kameshima later extended this process to include a separate active, ascorbic acid, which was then coated with oleate ions to provide an acid-resistant, ascorbic acid loaded delivery system.[25]

Similarly Trikeriotis encapsulated the hydrophobic antibiotics gramicidin and amphotericin B within cholate micelles which were then absorbed into the LDH structure *via* an ion exchange process.[26]

Li encapsulated prednisone in an amphiphilic block copolymer poly(*tert*-butyl acrylate-*co*-ethyl acrylate-*co*-methacrylic acid) (PTBEM) derived micelle which was intercalated into an LDH by a coprecipitation method.[27] The drug release rate increased in the order pH 4.8 < pH 6.4 < pH 7.2 which is opposite to that normally associated with non-micelle-LDH hybrid composites. This significant alteration in drug release behaviour was proposed to be due to the different specific release mechanisms at different pH values: Fickian diffusion (pH 7.2), anomalous transport (pH 6.4) and diffusion/erosion (pH 4.8) (Figures 7.7 and 7.8).

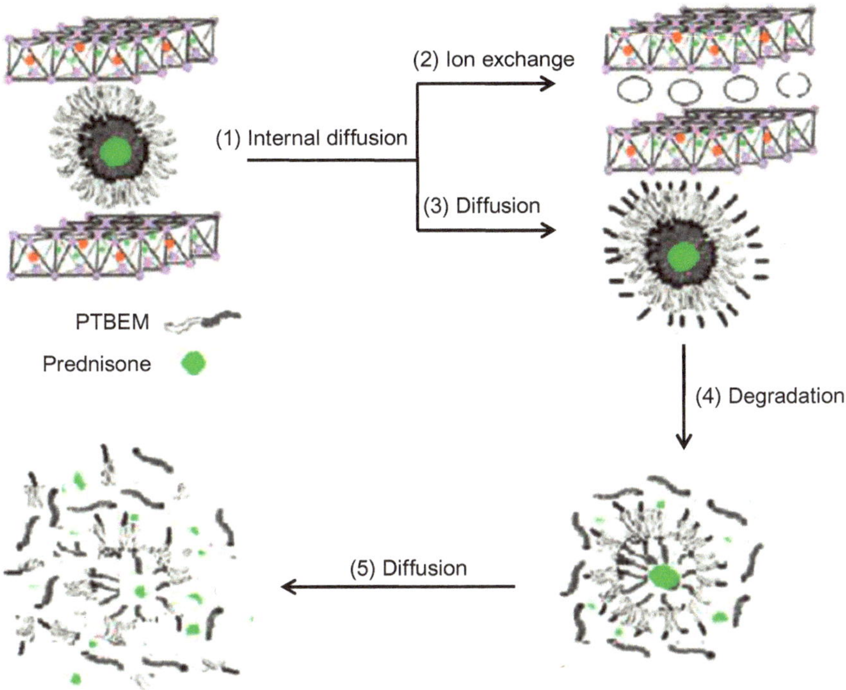

Figure 7.7 Schematic representation showing the different mechanisms involved during prednisone release from a prednisone-PTBEM-LDH nanocomposite as described by Li. *(Adapted from Ref. 27 with permission from Elsevier).*

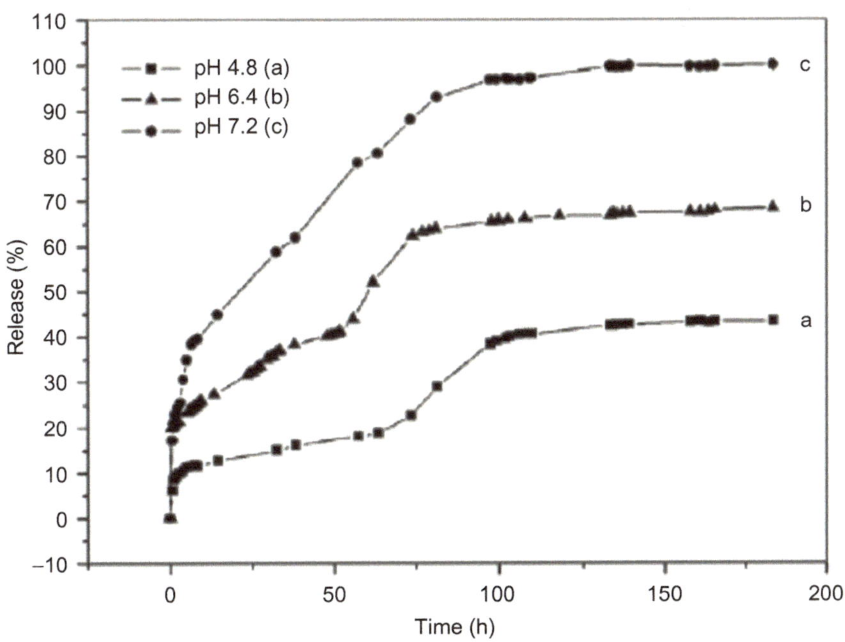

Figure 7.8 Prednisone release from prednisone-PTBEM-LDH drug delivery systems in different pH buffers. *(Reprinted from Ref. 27 with permission from Elsevier).*

7.2.1.3 Controlling Active Release from LDHs

Since the application of LDH materials for controlled drug delivery is a relatively new area of interest, systematic investigations into methods of controlling and modifying drug release from LDHs are still in their very early stages.

The method of drug-LDH conjugate preparation is very important. Intercalation of drug molecules by anion exchange and by the reconstruction method is usually carried out with large molar excesses of the incoming anion of interest in order to drive the equilibrium of the loading process through to completion. In this respect then it can be expected that the anionic layer space will be saturated with conjugate drug molecules. The extent of drug loading by the coprecipitation method is harder to predict. The required molar equivalent of drug is usually present in the desired stoichiometric amounts and is typically determined by equating it to one of the metal components. Full intercalation usually occurs but the presence of any other competing anions and how they might affect the LDH-drug composite forming process during coprecipitation should be considered. This effect can be clearly seen in Figure 7.9.[28] The three release profiles result from the three different active loading methods as already discussed. The material produced by the reconstruction method incorporates four times more methotrexate (42.3 wt%) than the other two methods, ion exchange and coprecipitation (9.9 and 10.4 wt%, respectively). Based on the reasons highlighted above (see Section 7.2.1.1), it is unsurprising that the reconstruction method allows a much higher incorporation of drug. The faster release of the drug incorporated by the reconstruction method can be explained on the grounds

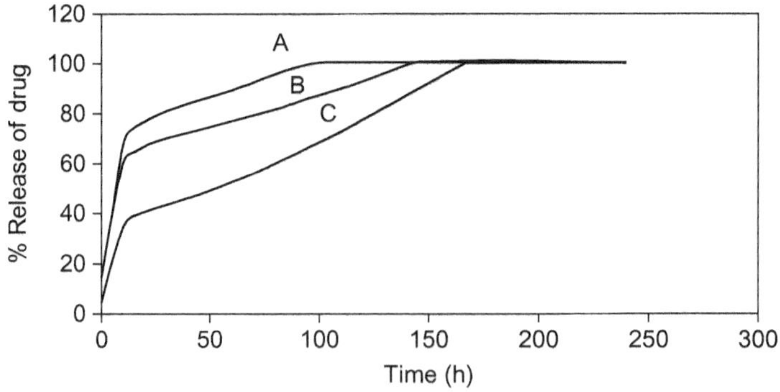

Figure 7.9 Comparative methotrexate release profiles from methotrexate-LDH composites prepared by the three different methods: A: reconstruction method; B: coprecipitation method and C: ion exchange method. *(Reprinted from Ref. 28 with permission from Elsevier).*

of a pure diffusion effect, a much higher diffusion gradient is produced by virtue of containing a much higher wt% of methotrexate to begin with. Samples B and C, coprecipitaion and ion exchange, respectively, contain equivalent amounts of drug, but the higher release rate from the coprecipitation sample (B) is possibly due to two effects:

(1) The presence of surface adsorbed drug following the coprecipitation method. This proportion of drug is released into the dissolution medium much quicker than any intercalated material.
(2) The presence of the drug during coprecipitation prevents effective crystal growth and as such results in smaller primary crystallites, as shown by laser diffraction determined particle size measurements. A smaller particle size for a diffusion-driven release system typically results in faster active release due to a much increased surface area and a smaller mean diffusion path length.

Particle morphology is also an important variable to consider. It has been shown that by carrying out the coprecipitation of LDHs at different pH levels the particle morphology can be influenced; at pH 7.2 worm-like or rod-like particles are formed whereas at pH 9.5 plate-like particles are formed.[29] It can be envisaged that the mean diffusion path length of a drug out of a particle with plate-like morphology will be shorter than a surface area equivalent particle with rod-like morphology, and as such rod-like particles will demonstrate more sustained release – based purely on diffusional mass transport considerations. As an extension of this effect, as with all diffusion controlled drug release systems, particle size will have an important effect on drug release. Kwak showed that by extending the ageing time of an LDH formed by coprecipitation that the mean particle size was increased (see Figure 7.10).[30] If this method could be extended to drug-loaded LDH drug delivery systems, the increased diffusion path from larger particles could be expected to have an effect on drug release rates.

A potentially more important parameter investigated by Pan was the changing Mg:Al ratio on drug release rates.[30] A decrease in the Mg:Al ratio, from 2.9 to 1.7, resulted in a decrease in the drug release rate and a more sustained-like release profile. This can be explained by a consideration of the charge density in the cationic layers with the changing Mg:Al ratio. As the relative amount of Al^{3+} is increased the overall positive charge of the hydroxide layer is also increased, resulting in a much stronger electrostatic attraction/ionic bonding interaction between the cationic hydroxide layer and the anionic intercalating species. Therefore provided that the release mechanism is governed by an ion exchange process only (as opposed

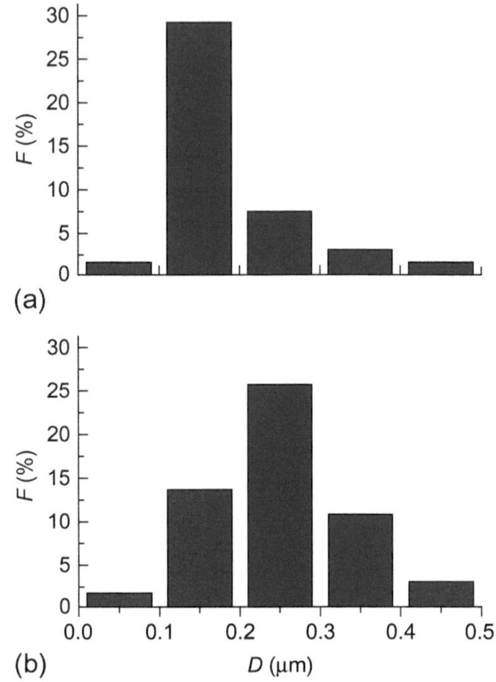

Figure 7.10 The change in particle size distribution with LDH ageing time: (a) 1 day and (b) 3 days. *(Adapted from Ref. 30 with permission from Elsevier).*

to the additional contribution of matrix dissolution at low pH), the energy barrier to overcome the more favourable matrix-drug interaction is going to be higher, resulting in a slower drug release.

A similar effect is seen when the monovalent Li^+ ion is included as a component of the hydroxide layer in place of Mg^{2+}, rather than altering the M^{2+}:M^{3+} ratio. The increased positive charge density of the Li/Al_2-LDH versus the Mg_2/Al-LDH has the effect of reducing the rate of fenbufen release due to a stronger drug-LDH interaction, as can be seen from the equivalent release profiles in Figure 7.11.[31]

Almost irrespective of the drug-LDH formulation properties, the *in vitro* method of evaluating the drug release from LDH materials is highly important in determining relative drug release kinetics. Since LDHs are basic in nature, at low pH, approximately less than 4, the hydroxide layers are dissolved and/or degraded. This has important implications for the design of LDH formulations for oral dosage in that any control of drug release will be lost once the low pH environment of the stomach is encountered.

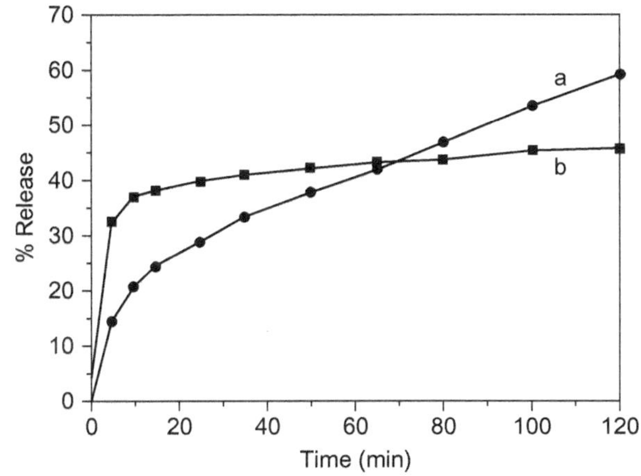

Figure 7.11 Fenbufen release from (a) Mg_2/Al-fenbufen-LDH and (b) Li/Al_2-fenbufen-LDH inorganic CRT systems. *(Reprinted from Ref. 31 with permission from Elsevier).*

In order to maintain controlled release the drug could be encapsulated within an acid-resistant micelle (which negates the purpose of the LDH as a carrier system), or the LDH formulation could have some form of acid-resistant coating, Eudragit® for example.[29] In conjunction with the pH of the test medium, the ionic strength is also very important since the second important drug-releasing mechanism relies on ionic exchange between the test medium and the LDH carrier system. Dissolution studies in various different phosphate buffers are common, and studies have shown that following partial or complete release of the intercalated drug species, the basal layer spacing is consistent with the presence of an anionic phosphate species. It can be seen therefore that the potential use of LDH drug delivery systems must be carefully considered and the *in vitro* tests carefully designed to emulate the *in vivo* conditions as closely as possible. In this way the drug-releasing properties of LDH formulations can be thoroughly assessed and evaluated in the context of the intended dosage form and application.

7.2.2 Cationic Clays

The vast majority of naturally found clays may be classified as cationic clays in that they consist of negatively charged structural layers in conjunction with cationic intercalating species. These materials are generally based on two-dimensional layered silicates. Structurally, silica tetrahedra and metal (M) octahedra are the basic building blocks of clay minerals, where typically $M = Al^{3+}$, Mg^{2+}, Fe^{2+} or Fe^{3+} and the distinct arrangement of these tetrahedra and octahedra gives rise to two broad families of clay materials. When

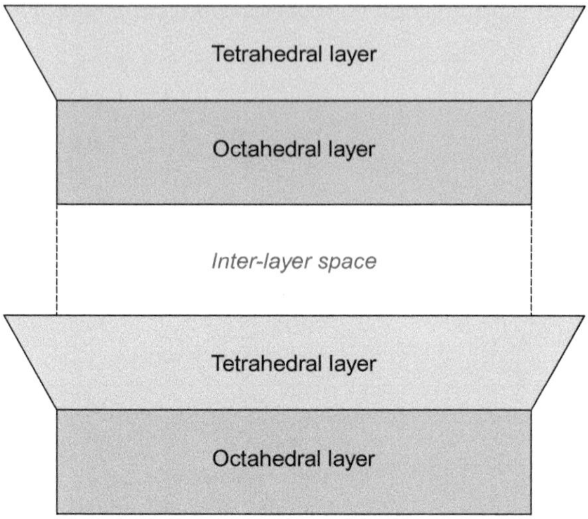

Figure 7.12 Diagrammatic representation of a 1:1 clay, including the interlayer space.

the structural layers consist of a tetrahedron-octahedron-tetrahedron a 2:1 clay material is formed, there are two tetrahedral layers for every one octahedral layer (see Figure 7.1). If the structural layers consist of a tetrahedral-octahedral arrangement, a system known as a 1:1 clay is formed (see Figure 7.12). As befits such a widely known and used natural mineral, the chemistry of clay minerals is vast and beyond the scope of the current text.[1,2]

Due to the structural layers of natural clays being negatively charged, these materials are complementary to LDHs in their ability to ion exchange cationic species. This opens up new opportunities for the intercalation and controlled delivery of a whole subset of pharmaceutical actives that are not suitable for formulation with LDH inorganic CRT systems. The net charge of natural clay systems can be formed by two separate phenomena[32]:

(1) Substitution of the Si^{4+} ions in the clay tetrahedra or Al^{3+} ions in the clay octahedra with ions of lower valency. For most common clay materials this is Al^{3+} or Mg^{2+}/Fe^{2+}, respectively. This substitution effect produces a net negative charge and is structural and thus permanent once the clay has been formed.
(2) Surface ionization states of non-bonding alumina (Al—OH) and silica (Si—OH) species. This charge effect is a dynamic situation which is dependent upon the local pH environment of the clay and the approximate isoelectric point of the hydroxyl groups present (see Ref. 33 for a detailed description of surface speciation and isoelectric points of some

common oxides). The important surface accessible hydroxyl groups typically reside on the basal surfaces of the tetrahedra in 2:1 clays, on the basal surfaces of the octahedra and tetrahedra in 1:1 clays, and along the sheet edges in all materials. In terms of the overall structure, this charge producing effect is negligible in most cases except for where ion substitution is zero.

7.2.2.1 Cationic Clays as inorganic CRT Systems

As with LDH inorganic CRT systems, the ion exchange potential and the structure and electrochemistry of the clay layers are the key elements for determining the ability of a clay material to act as a controlled release delivery system. In contrast to the LDHs, however, most cationic clays are only used as they are formed naturally. Unfortunately this limits the ability of the materials science researcher to build in modified functionality to these materials in order to tailor the drug-releasing properties. As has been discussed in Chapter 2 and Section 7.2.1, the drug-clay interactions are dominated by electrostatic interactions, in the case of cationic clays this interaction is determined by the drug molecule's cationic charge or hydrogen bonding potential and the anionic charge on the clay layers.

Cationic clays can be further classified into five separate groups of minerals as dictated by their structure and composition: smectites, vermiculites, illites, kaolinites and chlorites. As can be seen in Table 7.2, the surface area and ion exchange potential of the illite and chlorite groups of materials are very low, meaning that their active encapsulating potential is also very low. Whilst moderate external surface areas are available for active adsorption, the controlled release properties of such composites are limited. As such these materials are not discussed here for inorganic CRT applications any further.

In contrast, the vermiculite and smectite clays have exceptionally high surface areas and ion exchanging capacity (see Table 7.2).

Table 7.2 Typical ion exchange capacities and specific surface areas for different clays

Clay	Ion exchange capacities (mEq/100 g)	Surface area ($m^2\ g^{-1}$)		
		Internal	External	Total
Smectites	80-150	750	50	800
Vermiculites	120-200	750	<1	750
Illites	10-40	0	15	15
Kaolinites	1-10	0	15	15
Chlorites	<10	5	25	30

Adapted from Refs. 34,35.

It is on the basis of these two specific physical properties that smectites in particular have found initial success as inorganic CRT drug delivery systems, examples of which will now be highlighted.

7.2.2.2 Smectite Clays for inorganic CRT

Montmorillonite is by far the most well studied of the smectite clays for controlled release applications. Montmorillonite is differentiated by isomorphic substitution of Al^{3+} ions in the inner octahedral layer for Mg^{2+}. The resulting charge imbalance provides the structural net negative charge and the material's cation exchanging capacity. As there is no control over the formation of the clay structure, active intercalation can take place only by an ion exchange process. Sodium montmorillonite precursors are the most common for this purpose on similar grounds as for the LDH inorganic CRT systems, substitution of small, monovalent ions for larger multivalent ions is thermodynamically favoured since these ions are better able to satisfy the negative charge present within the clay structural layers.

Choy[36] used three different smectite clays as model carriers for the adsorption and *in vitro* release of donepezil, an acetylcholinesterase inhibitor used for the treatment of Alzheimer's disease. It was found that the drug intercalating capacity of the three different clays decreased in the order montmorillonite > saponite > laponite (see Table 7.3). Saponite and laponite clays are derived from isomorphic substitution of Al^{3+} ions for tetrahedral Si^{4+} ions and Fe^{2+} ions for octahedral Al^{3+} ions, respectively.

Increased cholinergic activity following donepezil dosage is known to cause excessive gastric acid secretion and as such gastrointestinal bleeding and occurrence of ulcers can be significant side effects. An added benefit of the clay inorganic CRT dosage form here is that clay materials are known to be excellent proton absorbers, so there is potential for the prevalence of gastric side effects to be reduced from these novel formulations. As a result, *in vitro* dissolution studies were carried out in pH 1.2 aqueous media. Unfortunately donepezil release in the first instance was found to be relatively slow

Table 7.3 Physical properties of clays used for controlled release of donepezil

Clay	Cation exchange capacity (mEq/100 g)	Drug loading (wt%)[a]
Montmorillonite	100	33.3-34.3
Saponite	71	22.0-23.3
Laponite	63	17.2-17.3

[a]Varies dependent on determination method, HPLC assay, or CHNS chemical analysis.

Table 7.4 Donepezil release under different *in vitro* conditions

Clay-drug composite	Donepezil release after 180 min in pH 1.2 HCl (%)	Donepezil release after 180 min in pH 1.2 HCl + Eudragit® E-100 (%)
Montmorillonite	7	12
Saponite	15	23
Laponite	37	61

and incomplete over a typical gastric residency time period (180 min), 37%, 15% and 7% release was seen for the laponite, saponite and montmorillonite formulations, respectively. It was thought that this effect was purely due to a correlation between the cation exchange capacity and the strength of the drug-clay ionic interaction.

In order to investigate the drug release mechanism and kinetics, the bulky, cationic polymethacrylate, copolymer Eudragit® E-100 was added to the low pH dissolution medium. Drug release for all three composites was increased, as seen in Table 7.4.

The Eudragit® E-100 copolymer is improving drug release by two separate mechanisms. The bulky cationic species will have a much greater preference to intercalate in the clay interlayer than the H^+ ions present in the dissolution medium and the donepezil already conjugated to the clay structure and as such will facilitate cationic exchange. This effect is most significant with the laponite composite where the ionic interaction between the laponite structural layers and the donepezil molecules is weakest. In addition, intercalation of the bulkier Eudragit® E-100 copolymer will increase the basal layer spacing relative to the donepezil intercalated material and as such the rate of drug diffusion out of the porous clay structure will be increased.

In the context of designing new cationic clay inorganic CRT systems, this study exemplified two important aspects:

- The cation exchange capacity of the substrate clay material is crucial in determining both the extent of active intercalation and also the rate of active release. Both the adsorption and desorption processes are fundamentally dependent on the nature of the ionic interactions between the guest active molecule and the host clay structure.
- The active release kinetics are determined by the mechanistic processes of (i) diffusion in of the exchanging cation, (ii) cation exchange and (iii) diffusion out of the active molecules.

These two aspects have important implications for considering the types of application cationic clay inorganic CRT systems are potentially useful for and how the necessary therapeutic drug release kinetics are going to be achieved.

Smectite clays have also been studied for the controlled release of antidepressants,[37] chemotherapeutics,[38,39] antibiotics,[40,41] ion channel blockers[42] and β-blockers.[37] In addition, the preparation of clay-polymer composite materials for improved formulations is also an active area of research.[43,44]

7.2.2.3 Other Clays for inorganic CRT

As can be seen from their typical properties in Table 7.2, kaolinite clays have very low surface area and cation exchanging capacity. Due to their unique 1:1 layer compared with the other clay groups, strong hydrogen bonds can be formed between the alumina and silica hydroxyl groups of the adjacent octahedral and tetrahedral layers, respectively. As such, kaolinite clays are often cited as being non-expansive. However, it has been shown that the intercalation of very small, highly polar molecules such as dimethyl sulphoxide,[45] *N*-methyl formamide,[46] methanol[47] and urea[48] breaks the intermolecular forces associated with the kaolinite layers allowing the intercalation of other more biologically useful molecules.[49,50]

Alternatively, halloysite a unique polymorph of kaolinite shows great promise for the encapsulation of active molecules. Due to the unique conditions present during its formation, rather than having a lamellar structure characteristic of kaolinite clays, halloysite has a nanotubular structure whereby the silica tetrahedral layer is located on the outside and the alumina octahedral layer is situated on the inside of the halloysite tubes (see Figure 7.13). As a consequence, halloysites have the unique ability to encapsulate anionic compounds within the tube pores and cationic compounds on the outer surface of the nanotubules. Unlike the structural charges present within the other clay materials, the net charge present on the internal and external halloysite surfaces is determined by the local pH environment. As such careful consideration is needed to enable loading of these materials with their desired actives. The first drug-loaded halloysite was reported by Price in 2001,[51] and the reader is referred to the excellent review articles of Lvov for an overview of the progress made in formulating drug-loaded halloysite inorganic CRT systems.[52,53]

Vermiculite clay is a 2:1 system having exceptionally high surface area and cation exchange capacity (see Table 7.2). Vermiculite clays are well known for their liquid adsorbing properties but only very recently has this fundamental physical property been investigated in the context of an inorganic CRT application.

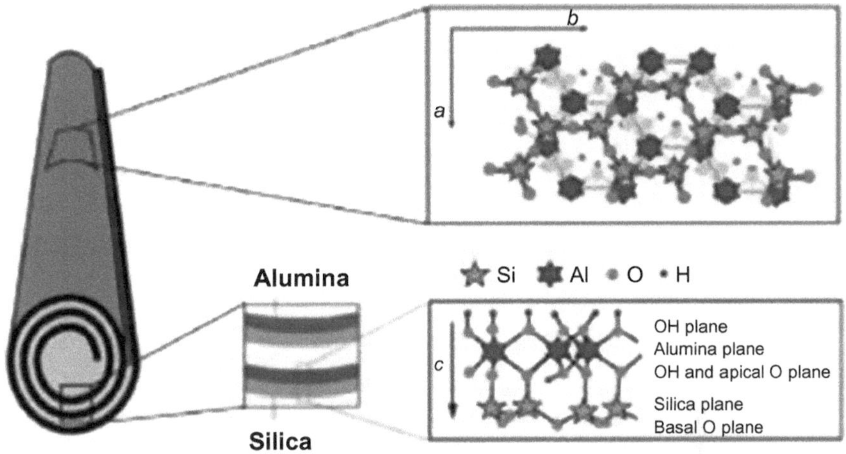

Figure 7.13 Schematic representation of a halloysite nanotube with its silica external surface and alumina lumen surface. *(Reprinted from Ref. 52 with permission from Elsevier).*

The ability of vermiculite to intercalate chlorhexidine diacetate for the purpose of preparing a local oral antibiotic has been investigated.[54] The drug-clay composite was found to have excellent antibacterial activity against *Escherichia coli, Enterococcus faecalis, and* in particular *Staphylococcus aureus*. These novel drug-clay composites were also tested for their *in vivo* toxicological properties following oral administration; neither local gastro-intestinal nor systemic reactions were observed showing that vermiculite is well tolerated *in vivo*.

7.3 ALTERNATIVE MESOPOROUS INORGANIC CRT MATERIALS

7.3.1 Mesoporous Zirconia

Mesoporous zirconia can be prepared in a similar manner to the mesoporous silicas and glasses mentioned in Chapters 4 and 5. Zirconium oxychloride or zirconium alkoxides are common precursors for the sol-gel process, though care must be taken when forming mixed oxide products since the zirconium alkoxides tend to rapidly hydrolyze under aqueous conditions. Often a zirconium chelating agent is added to prevent uncontrolled zirconium alkoxide hydrolysis and subsequent precipitation.[55] When added as a low mol% component of mesoporous bioactive glasses, the amorphous character is maintained, whilst more recently, methods have been developed to prepare metastable amorphous zirconia.[56] However, thermal treatment to remove the surfactant

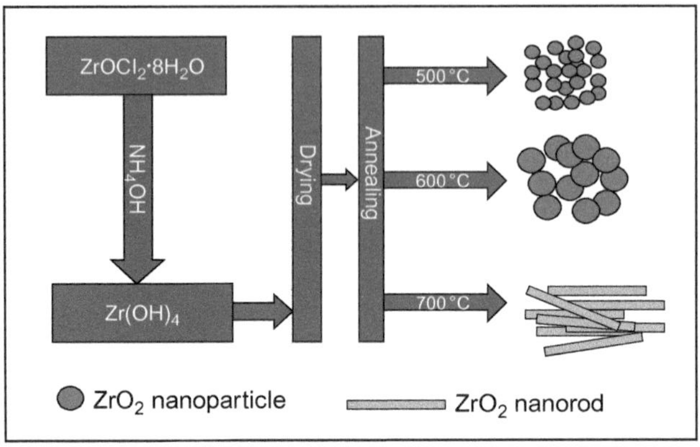

Figure 7.14 Formation of zirconia materials by a sol-gel method and how calcination temperature affects product morphology. *(Reprinted from Ref. 57 with permission from Elsevier).*

templating species tends to result in crystalline zirconia products.[57] Depending on the calcination temperature different particle morphologies can be formed (see Figure 7.14). At 500 °C 8 nm nanospheres are produced, at 600 °C 17 nm nanospheres are produced, and at 700 °C nanorods are produced.

Mesoporous zirconias in their own right have found limited use in drug delivery systems to date, but due to their extremely hard nature, lack of chemical reactivity and biocompatibility, zirconia is increasingly being used as a component material for coating metallic prostheses. Zirconia's radio-opacity has also meant that it has been investigated for use as a bone or dental cement.[55] Only very recently have these materials begun to be investigated for their drug-releasing properties. Indomethacin-loaded zirconia produced by a sol-gel method was used to dip-coat a titanium dental implant. The inclusion of polycaprolactone into this sol-gel preparation was also found to modify indomethacin release.[58]

Research into drug-releasing inorganic CRT zirconia materials is still in its very early stages, but considering the many excellent physical properties of zirconia, there is good potential for these materials to be used for bone and dental restoration products in particular.

7.3.2 Mesoporous Alumina

More recently, mesoporous alumina materials which are analogous to mesoporous silicas have been explored for controlled drug delivery.

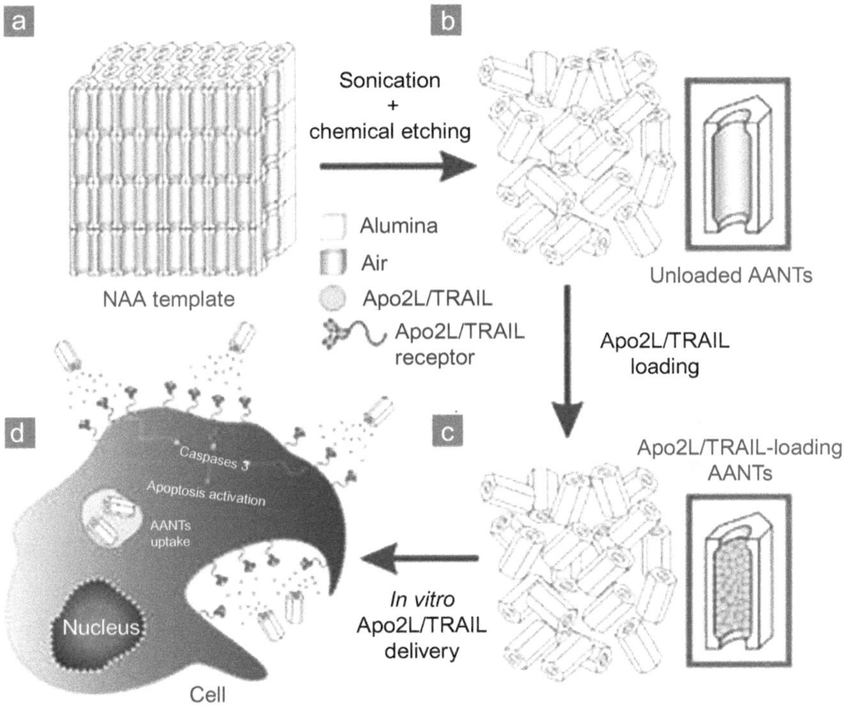

Figure 7.15 Schematic diagram showing drug loading and release of a cancer therapeutic by an anodized alumina. *(Reprinted from Ref. 59 with permission from Elsevier).*

Figure 7.16 Microscopy image showing typical nanoporous anodized alumina morpholgy. *(Reprinted from Ref. 59 with permission from Elsevier).*

These materials have been sourced in a number of different ways. Mesoporous aluminas are commercially available by an anodization process (see-Figures 7.15 and 7.16), by thermal decomposition of an aluminium nitrate salt, by a controlled precipitation, or by the Pyrosol method which can be considered as a low temperature chemical vapour deposition method, the pyrolysis of an aerosol. Porous aluminas can be made by sol-gel but this approach does not seem to have been well utilized for drug delivery to date.

As compared with the synthetic preparation of mesoporous silicas, the preparation of anodized alumina is a 'top-down' approach. These materials have a relatively high loading capacity compared with hydrogels, micro- and nanoparticles. It is thought that mesoporous alumina nanotubes have more accessible pores as compared with mesoporous silicas, as evidenced by the fact that the pre drug-loaded aluminas have relatively low surface area and pore volume, but the amount of drug that can be encapsulated within the pore volume is comparable with the corresponding mesoporous silica material.[60]

Porous aluminas have been investigated for the loading and delivery of ibuprofen,[61,62] naproxen,[60] and also for the potential of applying these materials to the surfaces of medical devices for further functionalization and drug loading.

As with silica-based mesoporous drug delivery systems, the alumina analogues have been studied for their potential to take place in surface modification reactions to (a) modulate the rate of release of an active or (b) increase the affinity of a molecule to be adsorbed to the modified alumina surface *via* more favourable electrostatic interactions.[62]

In many ways mesoporus alumina inorganic CRT materials are analogous to mesoporous silica drug delivery systems in that the main methodologies involved evolve around producing a well-ordered, non-soluble, porous matrix and then adsorbing the necessary active to the as-formed or modified alumina surface. In this respect, porous aluminas are complementary to porous silicas because at typical biological pH conditions, the alumina surface is positively charged as opposed to negatively charged with silica. In this way it can be seen that once the methods for applying porous aluminas as inorganic CRT materials are further developed then this change in surface charge means that the two materials are complementary to each other in terms of the types of drugs that can be more easily accommodated within the porous structure (Figure 7.17).

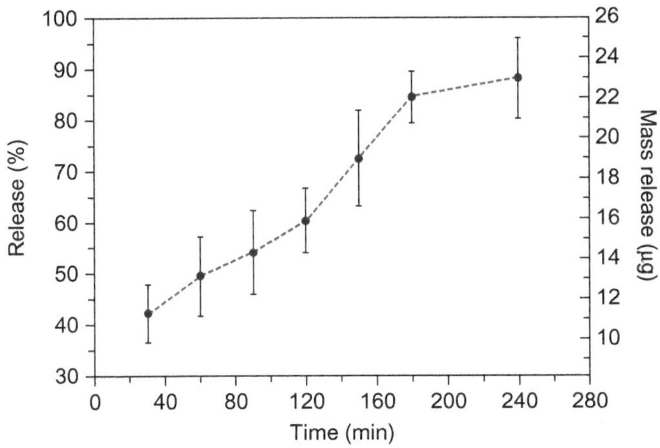

Figure 7.17 *In vitro* release of Apo2L/TRAIL from anodized amorphous nanotubes. *(Reprinted from Ref. 59 with permission from Elsevier).*

7.4 NANOPARTICLE DELIVERY SYSTEMS

Nanoparticles come in an almost infinite variety of sizes, shapes and functionalities, and modification and control of these properties has been an intense area of research since the beginning of nanotechnology. As a result there are a huge array of different materials and technologies involving the use of nanoparticles as drug delivery systems. It is not the intention to cover this topic intensively in this inorganic CRT book, it is out of the current scope and there are many reviews and reports on the topic where readers can find relevant information.[63–66] Here we use two examples, iron oxide and gold nanoparticles, to highlight some characteristics of nanoparticulate drug delivery systems.

7.4.1 Iron Oxide Nanoparticles

Despite the persistent toxicological concerns over the biological use of nanoparticles, the prevalence of work in understanding how to control nanoparticle features such as the individual particle size, the particle size distribution, their colloidal stability and surface charge, amongst many others, leads to the preparation of evermore useful and functional materials. This is especially the case in the development of functional particulate systems for biomedical applications.

Iron oxide nanoparticles (IONPs) are the most successful family of these materials for biomedical use. A number of IONP-based formulations for use as magnetic resonance imaging (MRI) contrast agents made it to late-stage development, and a few were approved for use, but only one product is currently available on the market, Lumirem (Guerbet) in Europe and Gastromark (Covidien) in the USA. This formulation is used to enhance delineation of the bowel to distinguish it from the organs and tissues adjacent to the upper gastrointestinal tract.[67]

In terms of the functional materials themselves, of the many forms of iron, only Fe_3O_4 and γ-Fe_2O_3 have found use for biomedical applications because of their favourable magnetic properties. IONP or superparamegnetic IONPs, as they are alternatively known, must be coated or surface-functionalized in order for their full theranostic potential to be realized.[68] IONP surface modifications:

(a) help to prevent degradation of the IONP core under *in vivo* conditions,
(b) prevent aggregation of the IONPs which would affect their functional magnetic properties, and
(c) provide surface functionality by which therapeutic agents can be attached.

In this way it can be seen that the nature of the surface functionalization imparts on IONP drug delivery systems biocompatibility, the necessary particle physical properties and the desired drug-releasing properties, whilst the nature and subsequent modification of the iron oxide core dictate the particle's magnetic properties.

The main methods of synthesizing IONPs are coprecipitation from mixtures of suitable metal precursors and the thermal decomposition of iron precursors either in the solid state or in organic solvents.

Due to their controllable nanosize and resulting superparamagnetism, IONPs have found particular use as contrast agents during MRI and for targeted drug delivery.

IONPs as contrast agents are typically administered either orally or intravenously. The specific nature of the surface modification for IONPs administered intravenously has to be very carefully designed – colloidal stability in the bloodstream, surface charge, nanoparticle size and non-specific blood protein binding are all important factors to consider. IONPs ranging in size from ~20 to 150 nm are most useful as they are known to accumulate in specific organ systems and thus make them useful for imaging in those tissues; the liver, spleen, kidneys, stomach and bones are all known accumulation sites for IONPs in this size range. Particles of this size tend to have prolonged residency in the bloodstream which increases their tendency to passively infiltrate solid tumours. Surface charge also plays an important role here. IONPs coated with neutral hydrophilic polymers are not actively targeted by blood proteins, whereas positively or negatively charged particles are, and as a result are quickly eliminated from the bloodstream. As well as passive tumour targeting, IONPs can also be used to actively target cancerous tissues by bonding known specific tumour biomarkers to the coated IONP surface.

This active tumour targeting behaviour as well as their ability to be guided by external magnetic fields makes IONPs excellent inorganic CRT candidates for delivery of chemotherapeutic agents to targeted tumour tissues. In terms of their mechanism of therapeutic effect, IONPs can be used to damage tumour cells *via* the production of a hyperthermia effect when placed in an AC electromagnetic field.

Additionally, IONPs can exert their therapeutic effect by the controlled release of chemotherapeutic drug molecules once the nanoparticles have reached their site of action. Drug molecules have been introduced into IONPs in a number of ways (see Figure 7.18).

Associated with these different drug-IONP conjugates are different strategies for drug release mechanisms. Doxorubicin is a common model compound for testing these drug delivery systems. Changes in temperature, pH and the presence of a magnetic field are all methods which have been used to release the drug compounds once the IONP conjugate has reached its desired site of action. The release of covalently bound drug molecules through the cleavage of labile hydrazone (acid hydrolysis), ester and disulphide (enzymatic cleavage) linkages has been particularly well studied.

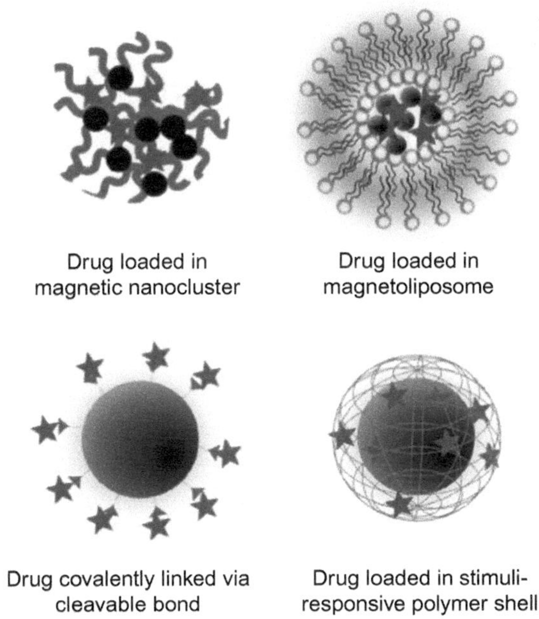

Figure 7.18 Schematic representation of different approaches to IONP-drug inorganic CRT systems. *(Reprinted from Ref. 69 with permission from Elsevier).*

7.4.2 Gold Nanoparticles

Gold nanoparticles have diverse optical and photothermal properties and their size can be finely tuned to modulate their toxicity and biodistribution. Gold nanoparticles have been used for diagnostic imaging[69,70] because of their ability to absorb near infrared and X-rays and their surface enhanced Raman scattering properties. More recently they have begun to be used for drug delivery[71–73] and more particularly photothermal treatment. Their ability to absorb near-infrared (NIR) energy and convert it to heat by a surface plasmon resonance[74] mechanism has made these materials promising for the treatment of solid tumours. A problem with gold nanoparticles is that the ideal particle size for efficient NIR absorption is 50–100 nm; however, smaller particles are required for effective accumulation in tumours. Smart nanoparticles 10 nm in size have been produced which are effective at entering tumour tissues and cells but under the lower pH conditions typical in tumour environments then aggregate to form larger nanoparticles.[75]

Gold nanoparticles are now being developed for drug delivery; hollow nanoparticles containing siRNA upon exposure to NIR changed their shape releasing the enclosed siRNA molecules (see Figure 7.19).[76] One of the major problems with gold nanoparticles is that they cannot be metabolized and tend to accumulate in the liver and spleen. One potential way of addressing this is by conjugating the gold nanoparticles with organic molecules. This methodology has been investigated by conjugating two different biological molecules to the surface of gold nanoparticles and evaluating the effect on the rate of clearance. It was found that glutathione-gold (Glu-Au) nanoparticles were effectively cleared through the kidneys whilst the bovine serum albumin-gold (BSA-Au) nanoparticles were not.[77]

The majority of gold nanoparticulate systems are prepared by the same general approach, reduction of gold salts to gold metal. Typically a stabilizing agent is required to be present during the reduction to prevent the aggregation of the nanoparticles as they are formed. The reduction of the precursor gold salts can be carried out chemically using a reducing agent such as sodium borohydride. Independent of the potential formation of aggregates following reduction, the chemical potential of the reducing agent, the amount present and the particular gold salt used, the solvent and reaction temperature are all important factors in determining the nanoparticle quality – the size morphology and uniformity of the product nanoparticles. Reduction *via* physical methods: microwaves, UV light, laser ablation, photochemical processes, ultrasound and heat are all common vectors for the reduction

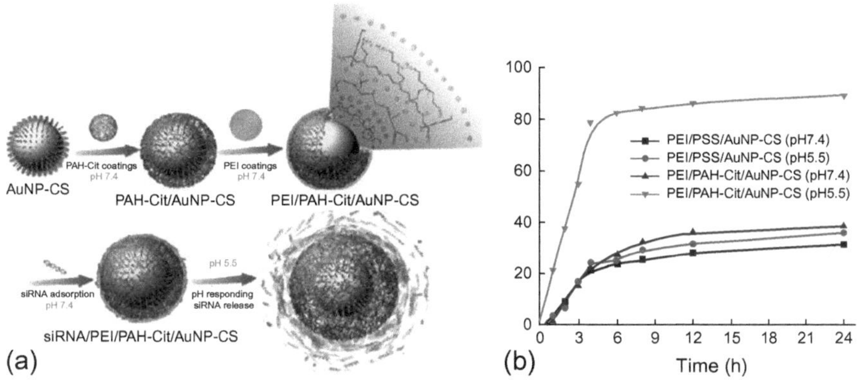

Figure 7.19 Schematic representation of (a) siRNA loading and desorption processes and (b) associated release siRNA release profiles. *(Adapted with permission from Ref. 76. Copyright 2012 American Chemical Society).*

process. Alternatively enzymatic reduction of gold salts has also been investigated.[78,79]

Gold nanoparticles can be readily used to adsorb biomolecules onto their surfaces, by definition, the huge available surface area present in nanoparticulate systems indicates that active loading capacities are high. The desorption process can be triggered by a number of external stimuli: changes in pH, changes in ionic strength, light cleavage of photo-labile surface active linking bonds and reduction of thiol links formed during chemisorption processes are all important methods of controlling the active desorption occurrence and rate.

In vivo, the presence of gold-thiol bonding interactions is important for allowing targeted delivery of conjugated active molecules. Gold surface functionalization with thiol groups can occur simultaneously during the nanoparticle forming process by having an organic thiol compound present during the gold salt reduction reaction. A functional thiol can then also be further functionalized. The desorption process can then be controlled by the presence of thiol groups in the local environment. An exchange process between the thiol groups bound to a gold surface with thiol groups in solution then results in active release. The presence of high levels of glutathione in certain cell types can be exploited to enable targeted release of a particular drug compound in those cells.[80]

As an example of their use as a drug delivery system, gold nanoparticles have been particularly well investigated for their ability to conjugate and then deliver DNA and RNA molecules.[81] Typically these large important biomolecules are physisorbed or covalently bound to the NP surface using thiol linkages. Before this conjugation process can take place though, a surface passivation process is usually required to either increase DNA/RNA-Au nanoparticle stabilization or endosomal release. A common strategy is the use of thiolated polymers (e.g., modified PEG, PAMA, PBAE) to firstly bind to the nanoparticle surface and then bond the active to the polymer. The gold-polymer linkage/interface and the polymer-active interface as well as the bulk properties of the polymer linking group are all possible mechanisms by which the active release can be controlled.

REFERENCES

1. Meurant G. *The chemistry of clay minerals*. Amsterdam: Elsevier; 2011.
2. Newman ACD. *Chemistry of clays and clay minerals*. Madison: Wiley; 1987.
3. Carretero MI, Pozo M. Clay and non-clay minerals in the pharmaceutical industry. *Appl Clay Sci* 2009;**46**(1):73–80.
4. Sorby D. Effect of adsorbents on drug absorption. I. Modification of promazine absorption by activated attapulgite and activated charcoal. *J Pharm Sci* 1965;**54**(5):677–83.
5. Sorby D, Liu G. Effects of adsorbents on drug absorption. II. Effect of an antidiarrhea mixture on promazine absorption. *J Pharm Sci* 1966;**55**(5):504–10.
6. de Roy A, Forano C, Besse JP. Layered double hydroxides: synthesis and post-synthesis modification. In: Rives V, editor. *Layered double hydroxides: present and future*. New York: Nova Science Publishers; 2001.
7. Costantino U, Nocchetti M. Layered double hydroxides and their intercalation compounds in photochemistry and in medicinal chemistry. In: Rives V, editor. *Layered double hydroxides: present and future*. New York: Nova Science Publishers; 2001.
8. Choy JH, Choi SJ, Oh JM, Park T. Clay minerals and layered double hydroxides for novel biological applications. *Appl Clay Sci* 2007;**36**:122–32.
9. Chakraborty M, Dasgupta S, Sengupta S, Chakraborty J, Basu D. Layered double hydroxides based ceramic nanocapsules as reservoir and carrier of functional anions. *Trans Ind Ceram Soc* 2010;**69**:153–63.
10. Ladewig K, Xu ZP, Lu GQ. Layered double hydroxide nanoparticles in gene and drug delivery. *Expert Opin Drug Deliv* 2009;**6**:907–22.
11. Wang Q, O'Hare D. Recent advances in the synthesis and application of layered double hydroxide (LDH) nanosheets. *Chem Rev* 2012;**112**:4124–55.
12. Bi X, Zhang H, Dou L. Layered double hydroxide-based nanocarriers for drug delivery. *Pharmaceutics* 2014;**6**:298–332.
13. Vaccari A. Preparation and catalytic properties of cationic and anionic clays. *Catal Today* 1998;**41**(1-3):53–71.
14. Cavani F, Trifirò F, Vaccari A. Hydrotalcite-type anionic clays: preparation, properties and applications. *Catal Today* 1991;**11**(2):173–301.
15. Yamaoka T, Abe M, Tsuji M. Synthesis of Cu/Al hydrotalcite like compound and its ion exchange property. *Mater Res Bull* 1989;**24**(10):1183–99.
16. Vicente R, del Arco M, Martín C. Intercalation of drugs in layered double hydroxides and their controlled release: a review. *Appl Clay Sci* 2014;**88-89**:239–69.
17. Choy JH, Kwak SY, Park JS, Jeong YJ, Portier J. Intercalative nanohybrids of nucleoside monophosphates and DNA in layered metal hydroxide. *J Am Chem Soc* 1999;**121**(6):1399–400.

18. Choy JH, Kwak SY, Jeong YJ, Park JS. Inorganic layered double hydroxides as nonviral vectors. *Angew Chem Int Ed* 2000;**39**(22):4041–5.
19. Choy JH, Park JS, Kwak SY, Jeong YJ, Han YS. Layered double hydroxide as gene reservoir. *Mol Cryst Liq Crys A* 2000;**341**(2):425–9.
20. Chibwe K, Jones W. Intercalation of organic and inorganic anions into layered double hydroxides. *J Chem Soc Chem Commun* 1989;**14**:926–7.
21. Kwon T, Pinnavaia TJ. Pillaring of a layered double hydroxide by polyoxometalates with Keggin-ion structures. *Chem Mater* 1989;**1**(4):381–3.
22. Ambrogi V, Fardella G, Grandolini G, Nocchetti M, Perioli L. Effect of hydrotalcite-like compounds on the aqueous solubility of some poorly water-soluble drugs. *J Pharm Sci* 2003;**92**(7):1407–18.
23. Tyner KM, Schiffman SR, Giannelis EP. Nanobiohybrids as delivery vehicles for camptothecin. *J Control Release* 2004;**95**(3):501–14.
24. Kameshima Y, Yoshizaki H, Nakajima A, Okada K. Preparation of sodium oleate/layered double hydroxide composites with acid-resistant properties. *J Colloid Interf Sci* 2006;**298**(2):624–8.
25. Kameshima Y, Sasaki H, Isobe T, Nakajima A, Okada K. Synthesis of composites of sodium oleate/Mg-Al-ascorbic acid-layered double hydroxides for drug delivery applications. *Int J Pharm* 2009;**381**(1):34–9.
26. Trikeriotis M, Ghanotakis DF. Intercalation of hydrophilic and hydrophobic antibiotics in layered double hydroxides. *Int J Pharm* 2007;**332**(1-2):176–84.
27. Li Y, Li H, Wei M, Lu J, Jin L. pH-responsive composite based on prednisone-block copolymer micelle intercalated inorganic layered matrix: structure and in vitro drug release. *Chem Eng J* 2009;**151**(1-3):359–66.
28. Chakraborty M, Dasgupta S, Bose P, Misra A, Kumar MT, Mitra M, et al. Layered double hydroxide: inorganic organic conjugate nanocarrier for methotrexate. *J Phys Chem Solids* 2011;**72**(6):779–83.
29. Pan D, Zhang H, Zhang T, Duan X. A novel organic-inorganic microhybrids containing anticancer agent doxifluridine and layered double hydroxides: structure and controlled release properties. *Chem Eng Sci* 2010;**65**(12):3762–71.
30. Kwak SY, Kriven WM, Wallig MA, Choy JH. Inorganic delivery vector for intravenous injection. *Biomaterials* 2004;**25**(28):5995–6001.
31. Li B, He J, Gevans D, Duan X. Inorganic layered double hydroxides as a drug delivery system – intercalation and in vitro release of fenbufen. *Appl Clay* 2004;**27**(3-4):199–207.
32. Eslinger E, Pevear DR. *Clay minerals for petroleum geologists and engineers*. SEPM: University of Michigan; 1988.
33. Kosmulski M. *Chemical properties of materials surfaces*. New York: Marcel Dekker; 2001.
34. van Olphen H, Fripiat JJ. *Data handbook for clay materials and other non-metallic minerals*. New York: Pergamon Press; 1979, p. 346.
35. Drever JI. *The geochemistry of natural waters*. Englewood Cliffs, NJ: Prentice-Hall Inc.; 1982, p. 82.
36. Park JK, Choy YB, Oh JM, Kim JY, Hwang SJ, Choy JH. Controlled release of donepezil intercalated in smectite clays. *Int J Pharm* 2008;**359**(1-2):198–204.
37. Joshi GV, Kevadiya BD, Patel HA, Bajaj HC, Jasra RV. Montmorillonite as a drug delivery system: intercalation and in vitro release of timolol maleate. *Int J Pharm* 2009;**374**(1-2):53–7.
38. Apps MG, Ammit AJ, Gu A, Wheate NJ. Analysis of montmorillonite clay as a vehicle in platinum anticancer drug delivery. *Inorg Chim Acta* 2014;**421**:513–8.
39. Iliescu RI, Andronescu E, Voicu G, Ficai A, Covaliu CI. Hybrid materials based on montmorillonite and citostatic drugs: preparation and characterization. *Appl Clay Sci* 2011;**52**(1-2):62–8.

40. Calabrese I, Cavallaro G, Scialabba C, Licciardi M, Merli M, Sciascia L, et al. Montmorillonitenanodevices for the colon metronidazole delivery. *Int J Pharm* 2013;**457**(1):224–36.
41. Hamilton AR, Hutcheon GA, Roberts M, Gaskell EE. Formulation and antibacterial profiles of clay-ciprofloxacin composites. *Appl Clay Sci* 2014;**87**:129–35.
42. Marques IJ, Vaz PD, Fernandes AC, Nunes CD. Advantageous delivery of nifedipine from inorganic materials showing increased solubility and biocompatibility. *Microporous Mesoporous Mater* 2014;**183**:192–200.
43. Jain S, Datta M. Oral extended release of dexamethasone: montmorillonite-PLGA nanocomposites as a delivery vehicle. *Appl Clay Sci* 2014;**104**:182–8.
44. Kevadiya BD, Rajkumar S, Bajaj HC. Application and evaluation of layered silicate-chitosan composites for site specific delivery of diclofenac. *Biocybern Biomed Eng* 2014; http://dx.doi.org/10.1016/j.bbe.2014.08.004.
45. Olejnik S, Posner AM, Quirk JP. The IR spectra ofinterlamellar kaolinite-amide complexes. I. The complexes of formamide, N-methylformamide and dimethylformamide. *Clay Clay Miner* 1971;**19**:83.
46. Adams JM. *Clay Clay Miner* 1978;**26**:169–72.
47. Komori Y, Sugahara Y, Kuroda K. Intercalation of alkylamines and water into kaolinite with methanol kaolinite as an intermediate. *Appl Clay Sci* 1999;**15**(1-2):241–52.
48. Caglar B. Structural characterization of kaolinite-nicotinamide intercalation composite. *J Mol Struct* 2012;**1020**:48–55.
49. Matusik J, Gaweł A, Bahranowski K. Grafting of methanol in dickite and intercalation of hexylamine. *Appl Clay Sci* 2012;**56**:63–7.
50. Yuan P, Tan D, Annabi-Bergaya F, Yan W, Liu D, Liu Z. From platy kaolinite to aluminosilicatenanoroll via one-step delamination of kaolinite: effect of the temperature of intercalation. *Appl Clay Sci* 2013;**83-84**:68–76.
51. Price R, Lvov Y, Gaber BP. In vitro release characteristics of tetracycline HCl, khellin and nicotinamide adenine dineculeotide from halloysite; a cylindrical mineral. *J Microencapsul* 2001;**18**(6):713–22.
52. Lvov Y, Abdullayev E. Functional polymer-clay nanotube composites with sustained release of chemical agents. *Prog Polm Sci* 2013;**38**(10-11):1690–719.
53. Shutava TG, Fakhrullin RF, Lvov YM. Spherical and tubule nanocarriers for sustained drug release. *Curr Opin Pharmacol* 2014;**18**:141–8.
54. Holešová S, Stembírek J, Bartošová L, Pražanová G, Valášková M, Samlíková M, et al. Antibacterial efficiency of vermiculite/chlorhexidinenanocomposites and results of the in vivo test of harmlessness of vermiculite. *Mater Sci Eng C Biol* 2014;**42**:466–73.
55. Tallia F, Gallo M, Pontiroli L, Baino F, Fiorilli S, Onida B, et al. Zirconia-containing radiopaque mesoporous bioactive glasses. *Mater Lett* 2014;**130**:281–4.
56. Raghavendra VB, Naik S, Antony M, Ramalingam G, Rajamathi M, Raghavan S. Amorphous, monoclinic, and tetragonal porous zirconia through a controlled self-sustained combustion route. *J Am Ceram Soc* 2011;**94**(6):1747–55.
57. Kumar S, Bhunia S, Ojha AK. Effect of calcination temperature on phase transformation, structural and optical properties of sol-gel derived ZrO_2 nanostructures. *Physica E* 2015;**66**:74–80.
58. Catauro M, Bollino F, Papale F, Pacifico S. Modulation of indomethacin release from ZrO_2/PCL hybrid multilayers synthesized via sol-gel dip coating. *J Drug Deliv Sci Technol* 2015;**26**:10–6.
59. Wang Y, Santos A, Kaur G, Evdokiou A, Losic D. Structurally engineered anodic alumina nanotubes as nano-carriers for delivery of anticancer therapeutics. *Biomaterials* 2014;**35**(21):5517–26.
60. Carriazo D, del Arco M, Martín C, Ramos C, Rives V. Influence of the inorganic matrix nature on the sustained release of naproxen. *Microporous Mesoporous Mater* 2010;**130**(1-3):229–38.

61. Granado S, Ragel V, Cabanas V, San Roman J, Vallet-Regi M. Influence of α-Al_2O_3 morphology and particle size on drug release from ceramic/polymer composites. *J Mater Chem* 1997;**7**(8):1581–5.
62. Kapoor S, Hegde R, Bhattacharyya AJ. Influence of surface chemistry of mesoporous alumina with wide pore distribution on controlled drug release. *J Control Release* 2009;**140**(1):34–9.
63. Svenson S, Prud'homme RK, editors. *Multifunctional nanoparticles for drug delivery applications: imaging, targeting, and delivery*. New York: Springer; 2012.
64. Gupta RB, Kompella UB. *Nanoparticle technology for drug delivery*. Abingdon: Taylor & Francis; 2006.
65. Thassu D, Deleers M, Pathak YV, editors. *Nanoparticulate drug delivery systems*. Boca Raton: Taylor & Francis Group; 2007.
66. de Villiers MM, Aramwit P, Kwon GS, editors. *Nanotechnology in drug delivery*. New York: Springer; 2008.
67. GastroMARK patient information leaflet, www.amagpharma.com/documents/products/pdfs/gastromark_insert.pdf, MKR 1120103 revised 01/03 [accessed 12.04.15].
68. Hola K, Markova Z, Zoppellaro G, Tucek J, Zboril R. Tailored functionalization of iron oxide nanoparticles for MRI, drug delivery, magnetic separation and immobilization of biosubstances. *Biotechnol Adv* 2015;http://dx.doi.org/10.1016/j.biotechadv.2015.02.003.
69. Lai SF, Chien CC, Chen WC, Chen HH, Chen YY, Wang CL, et al. Very small photoluminescent gold nanoparticles for multimodality biomedical imaging. *Biotechnol Adv* 2013;**31**(3):362–8.
70. Cho EC, Glaus C, Chen J, Welch MJ, Xia Y. Inorganic nanoparticle-based contrast agents for molecular imaging. *Trends Mol Med* 2010;**16**(12):561–73.
71. Vigderman L, Zubarev ER. Therapeutic platforms based on gold nanoparticles and their covalent conjugates with drug molecules. *Adv Drug Deliv Rev* 2013;**65**(5):663–76.
72. Ajnai G, Chiu A, Kan T, Cheng CC, Tsai TH, Chang J. Trends of gold nanoparticle-based drug delivery system in cancer therapy. *J Exp Clin Med* 2014;**6**(6):172–8.
73. Spencer DS, Puranik AS, Peppas NA. Intelligent nanoparticles for advanced drug delivery in cancer treatment. *Curr Opin Chem Eng* 2015;**7**:84–92.
74. Cao J, Sun T, Grattan KTV. Gold nanorod-based localized surface plasmon resonance biosensors: a review. *Sens Actuator B Chem* 2014;**195**:332–51.
75. Nam J, La WG, Hwang S, Ha YS, Park N, Won N, et al. pH-responsive assembly of gold nanoparticles and 'spatiotemporally concerted' drug release for synergistic cancer therapy. *ACS Nano* 2013;**7**(4):3388–402.
76. Han L, Zhao J, Zhang X, Cao W, Hu X, Zou G, et al. Enhanced siRNA delivery and silencing gold-chitosan nanosystem with surface charge-reversal polymer assembly and good biocompatibility. *ACS Nano* 2012;**6**(8):7340–51.
77. Urie R, Rege K. Nanoscale inorganic scaffolds as therapeutics and delivery vehicles. *Curr Opin Chem Eng* 2015;**7**:120–8.
78. Khan MS, Vishakante GD, Siddaramaiah H. Gold nanoparticles: a paradigm shift in biomedical applications. *Adv Colloid Interf* 2013;**199-200**:44–58.
79. Zhao P, Li N, Astruc D. State of the art in gold nanoparticle synthesis. *Coord Chem Rev* 2013;**257**(3-4):638–65.
80. Hong R, Han G, Fernández JM, Kim BJ, Forbes NS, Rotello VM. Glutathione-mediated delivery and release using monolayer protected nanoparticle carriers. *J Am Chem Soc* 2006;**128**(4):1078–9.
81. Malmsten M. Inorganic nanomaterials as delivery systems for proteins, peptides, DNA, and siRNA. *Curr Opin Colloid Interf Sci* 2013;**18**(5):468–80.

CHAPTER 8

Future Development of iCRT

Contents

8.1 Overview 222
8.2 Design and Development of Inorganic CRT Systems 222
8.2.1 Development of new inorganic CRT thermodynamic theories on solubility 222
8.2.2 Development of new inorganic CRT kinetic theories for controlled release 226
8.2.3 Design and development of an inorganic CRT system for drug controlled release 228
8.2.3.1 Design and development of inorganic CRT oral formulations for poorly soluble drugs 230
8.2.3.2 Design and development of inorganic CRT oral formulations for very soluble drugs 231
8.2.3.3 Design and development of inorganic CRT for abuse-deterrent application 232
8.2.3.4 Design and development of inorganic CRT for protein therapeutics 232
8.3 Regulation and Relevant Issues 233
8.3.1 Introduction 233
8.3.2 Design control for product development 234
8.3.2.1 Design and development planning 235
8.3.2.2 Design input 235
8.3.2.3 Design output 236
8.3.2.4 Design review 236
8.3.2.5 Design verification/validation 237
8.3.2.6 Design transfer 238
8.3.2.7 Design change 238
8.3.2.8 Design history file 238
8.3.3 Note for design drug development 239
8.3.3.1 Drug substance 239
8.3.3.2 Excipients 239
8.3.3.3 Formulation development 240
8.3.3.4 Physicochemical and biological properties 241
8.3.3.5 Microbiological attributes 241
8.3.3.6 Compatibility 241
8.4 Summary 241
References 242
Further reading 242

Inorganic Controlled Release Technology
http://dx.doi.org/10.1016/B978-0-08-099991-3.00008-9

8.1 OVERVIEW

From a materials science and technology point of view, inorganic controlled release technology (inorganic CRT) is one of the new members of the nanotechnology family. It is no doubt a new and exciting field that offers scientists the opportunity to control matter at very small dimensions, opening up many possibilities for developing a variety of new products for the controlled release of drugs or other active ingredients.

The original definition of nanotechnology often refers to structural sizes roughly in the 1-100 nm size range in at least one dimension. With respect to inorganic CRT, we aim to work in the region at the lower end of this definition, that is the structural size is down to a few nanometres, or tens of nanometres at the most extreme circumstances, as discussed in Chapters 1-7 in this book. For an inorganic CRT system that serves as a drug carrier, we aim to accommodate the drug in such a way that will prevent drug molecules from forming 'rigid' crystal structures, but to keep them in a 'soft' and more controllable amorphous state (refer to Chapter 1 for the crystal and amorphous definitions). By doing so, drug molecules can be released individually so they will become active much faster than with a classical drug delivery formulation technology where a drug (most of which are in crystal forms) has to be dissolved first before entering into the blood circulation system.

As a new branch of science and technology, we are going to highlight some areas worth paying attention to for future development in view of design, development, and relevant regulatory concerns.

8.2 DESIGN AND DEVELOPMENT OF INORGANIC CRT SYSTEMS

Before discussing the topic of 'Design and Development of inorganic CRT Systems', let us firstly review relevant theories that have been developed to date on inorganic CRT science and technology.

8.2.1 Development of New Inorganic CRT Thermodynamic Theories on Solubility

The basic thermodynamic relationship governing solubility is given by Gibbs free energy ΔG_s. The relationship between solubility, Gibbs free energy ΔG_s and a drug dissolution enthalpy ΔH_s and entropy ΔS_s at temperature T is given by

$$\Delta G_s = \Delta H_s - T\Delta S_s \tag{8.1}$$

For a system consisting of a drug assigned as 'a', a solvent as 'b' and inorganic CRT system as 'i', the thermodynamic enthalpy and entropy components of Equation (8.1) are given as follows:

$$\Delta H_s = \Delta H_{aa} + \Delta H_{ab} + \Delta H_{ai} + \Delta H_{bi} + \Delta H_{bb} \tag{8.2}$$

$$\Delta S_s = \Delta S_{aa} + \Delta S_{ab} + \Delta S_{ai} + \Delta S_{bi} + \Delta S_{bb} \tag{8.3}$$

where enthalpy ΔH_{aa} is a drug cohesive energy change, say disruption of the crystal lattice to individual drug molecules, ΔH_{ab} is the interaction energy between drug molecules and solvent molecules, ΔH_{ai} is the interaction energy between drug molecules and the inorganic CRT system, ΔH_{bi} is the interaction energy between the solvent and the inorganic CRT system and ΔH_{bb} is the enthalpy of the inorganic CRT, represented by the energy change associated with a phase change in the inorganic CRT system if applicable (e.g. dissolution of the carrier system). The corresponding entropies are given as ΔS_{aa}, ΔS_{ab}, ΔS_{ai}, ΔS_{bi} and ΔS_{bb}, respectively.

For classical drug dissolution without inorganic CRT: Equations (8.2) and (8.3) become

$$\Delta H_s = \Delta H_{aa} + \Delta H_{ab} \tag{8.4}$$

$$\Delta S_s = \Delta S_{aa} + \Delta S_{ab} \tag{8.5}$$

We will consider two extreme examples of drug compounds: the first is a very water-soluble entity, while the second is a poorly water-soluble compound. These are two of the most challenging subjects in terms of producing drug controlled release formulations. This is particularly the case for the latter as the number of poorly water-soluble drug candidates has dramatically increased recently. Any technology that will lead to solutions in this area of formulation science is welcome.

For a very water-soluble drug, enthalpy ΔH_s is normally not a significant component, while the energetic contribution of the enthalpy change ΔS_s is generally large. This combined effect leads to the drug compound being easily dissolved in an appropriate condition. In contrast, for a poorly water-soluble drug compound, the drug molecules are normally found in the crystalline state in which case great energy is required to be input into the system to overcome the crystal lattice energy ΔH_{aa}. Only after that it is possible for the drug molecules to be dissolved and become solvated. The poorly water-soluble crystal also means that the contribution of ΔH_{aa} to the Gibbs free energy change, ΔG_s, is high. Before dissolution into the solvent ΔS_{aa} and ΔS_{ab} are close to zero. This, plus the high interaction energy between the poorly soluble drug and water, usually leads to $\Delta H_{ab} > 0$ and then $\Delta G_s > 0$ in most circumstances. So from a theoretical consideration it is not possible to dissolve poorly soluble compounds in water, or they have very limited rates of dissolution. The only way forward is to artificially reduce ΔH_s and increase ΔS_s. One common strategy to this approach is by milling drug particles/agglomerates as small as possible, although there will always be limitations to the amount of size reduction that can take place with any one particular drug compound.

For inorganic CRT drug systems: With reference to the models given in Figures 2.13 and 2.19 the mesoporous nanopore sizes are in the range of a few nanometres (also refer to experimental data of MCM-41 in Chapters 3-5). The important fact is that a drug compound is dispersed and distributed to a pseudo-molecular level into the inorganic CRT system, and no crystal lattice energy needs to be overcome, that is $\Delta H_{aa} = 0$. This is because the energy needed to separate one molecule from a crystalline phase is no longer required – in which case, $\Delta S_{aa} = 0$ too. Equations (8.2) and (8.3) are then given as

$$\Delta H_s = \Delta H_{ab} + \Delta H_{ai} + \Delta H_{bi} + \Delta H_{bb} \tag{8.6}$$

$$\Delta S_s = \Delta S_{ab} + \Delta S_{ai} + \Delta S_{bi} + \Delta S_{bb} \tag{8.7}$$

Without crystal enthalpy for a poorly soluble drug, the probability for $\Delta G_s < 0$ is much higher in that we do not need to spend energy to separate individual drug molecules from their crystal phase, which is the most energy consuming process in the dissolution process. This means that an inorganic CRT system can remove one major hurdle in making poorly soluble drugs soluble. From a thermodynamical perspective an inorganic CRT system has a greater tendency to enable a poorly soluble drug to dissolve as soon as it encounters, and is able to interact with, solvent molecules.

Equations (8.6) and (8.7) are applicable to an inorganic CRT system that is water soluble. In the case where an *inorganic CRT system is not water soluble*, the enthalpy and entropy associated with the inorganic CRT carrier itself make very little contribution to the total Gibbs energy. So with $\Delta H_{bb} = 0$, $\Delta S_{bb} = 0$, Equations (8.2) and (8.3) become:

$$\Delta H_s = \Delta H_{ab} + \Delta H_{ai} + \Delta H_{bi} \tag{8.8}$$

$$\Delta S_s = \Delta S_{ab} + \Delta S_{ai} + \Delta S_{bi} \tag{8.9}$$

Up to this point, only ΔH_{ab} and ΔS_{ab} in the equations above have solutions which have been discussed in Section 4.3.1 in Chapter 4. Following the same logic, both enthalpy and entropy can be worked out in the same way:

The dissolution entropy change ΔS_{ab}:

$$\Delta S_{ab} = -\kappa\left(\frac{\phi_a}{N}\ln\phi_a + \frac{\phi_b}{N}\ln\phi_b\right) \tag{8.10}$$

and the dissolution enthalpy change ΔH_{ab}:

$$\Delta H_{ab} = \kappa T\phi_A\phi_B\chi_{AB} \tag{8.11}$$

Here we still use interaction parameter, χ_{ab} defined by Flory-Huggins:

$$\chi_{ab} = \left(\frac{z}{N_B T}\right)\left\{\varepsilon_{ab} - \frac{1}{2}(\varepsilon_{aa} + \varepsilon_{bb})\right\} \tag{8.12}$$

where κ is the Boltzmann constant, $N = (N_a + N_b)$ the number of interaction units between a solvent (such as water) and the drug molecules, ϕ_a and ϕ_b are the volume fractions of drug 'a' and solvent 'b', z is the coordination number (the number of nearest neighbours surrounding a given interaction unit), and ε_{ab}, ε_{aa} and ε_{bb} are the interaction energies per interaction per unit between 'a-b', 'a-a' and 'b-b' in a solution.

Moving the theory a step further, we assume that in the liquid state, $\varepsilon_{bb} \approx 0$, that is the exchanging of adjacent water molecules in a local environment involves no net energy change. Equation (8.12) then becomes

$$\chi_{ab} = \left(\frac{z}{N_B T}\right)\left\{\varepsilon_{ab} - \frac{\varepsilon_{aa}}{2}\right\} \tag{8.13}$$

Equations (8.12) and (8.13) can be treated as general equations that account for the molecular interactions between a pair of materials 'a' and 'b', which are applicable to all materials involved in the dissolution process of the inorganic CRT system. Even so, there is still much theoretical and experimental work yet to be completed in order to develop solutions for Equations (8.6) and (8.7) and to obtain physical parameters to give a complete solution of the dissolution thermodynamics using inorganic CRT systems. These thermodynamic theories and dissolution kinetics together can help researchers understand the factors that affect inorganic CRT controlled release mechanisms and to aid design of inorganic CRT systems for specific applications in the future. This is particularly important for the water-soluble inorganic CRT system.

The above thermodynamics are also applicable to the inorganic CRT drug-loading process. All the above dissolution equations apply to the process of drug loading into a mesoporous inorganic CRT system. The working principles and the thermodynamic theories are the same for both. It is addressed here that in both drug-loading and releasing processes, molecular interaction between the drug/s, solvent/s and inorganic CRT system is the key to success for new drug development using the inorganic CRT system.

8.2.2 Development of New Inorganic CRT Kinetic Theories for Controlled Release

From a design point of view, Chapter 4 has set a general foundation for both templated and non-templated silicate-based inorganic CRT systems. However, all the technology discussed is mainly focussed on how to form a porous carrier at the start without considering the drug structure, interactions and drug functionality required for therapeutic action. A better inorganic CRT system should at least take the drug chemistry and the material physics into account before formulation development. These factors should aid in the determination of which kind of inorganic CRT system is to be designed and developed. It must also be such a system that can meet the clinical requirements necessary to maximize the efficacy of the drug for its intended use.

The fundamental science and technology of inorganic CRT systems are still at an early stage of development, and there is still a big gap between understanding the processes and properties for improving controlled drug release and using this knowledge to develop market-ready products. More education and research are necessary to develop the existing technology further and to exploit application areas where inorganic CRT formulations could provide improved controlled release that are as yet unexplored. Its ultimate goal is for both healthcare and non-healthcare applications.

In Chapter 2, we used a simple one-dimensional model to work out the most popular kinetic equations that have been used by many researchers in the field. However, the complexity of the kinetics of mass transportation and further development of those theories are beyond the scope of this book. Here we recommend interested professionals to work on these theories as there are several questions that need better solutions for both theoretical exploitation and experimental verification of the inorganic CRT system as far as the working mechanism is concerned for controlled release of active ingredients.

The molecular diffusion model shown in Figure 2.13 is a simplified model for a drug-loaded mesoporous inorganic CRT system. For future development, we should consider many factors that will affect the kinetics of an inorganic CRT system for drug controlled release. The new theories should at least consider the following factors when studying inorganic CRT kinetics. Initially, when a solvent such as water is brought into contact with an inorganic CRT drug-loaded system, several mass transfer processes will occur, which at least include

(a) firstly, a nanopore in an inorganic CRT system will be filled with water; this is a diffusion controlled process;
(b) secondly, drug molecules will be dissolved in the nanopores *via* interaction with water molecules; this is a dissolution process and
(c) the dissolved drug molecules will diffuse out of the nanopore; this is also a diffusion controlled process.

The situation is not quite as straightforward as is suggested. Complexity is added if the working mechanisms are a combination of the actions of (a), (b) and (c), the action speed of which varies from time to time. These at least include the following possible combinations:

(1-1) (a) > (b)
(1-2) (a) = (b)
(1-3) (a) < (b)
(2.1) (b) > (c)
(2.2) (b) = (c)
(2.3) (b) < (c)
(3.1) (c) > (a)
(3.2) (c) = (a)
(3.3) (c) < (a)

These kinetic variations will affect overall drug releasing kinetics.

Evaluation of the drug releasing kinetics becomes more complicated again when the inorganic CRT system is partially or wholly water soluble. In this instance, the diffusion theories discussed in Chapter 2 will not be adequate enough to work through more complicated situations like these combinations. A dissolving inorganic CRT carrier will add more variables/factors to the above nine circumstances. So the theories and verification of the theories by experiments are far behind the progress made to date of the inorganic CRT technology. Much more education and research of the inorganic CRT technology are needed to more fully understand the fundamental physical processes controlling drug release in order to expedite the development of this new science and technology for different applications.

The kinetic theories discussed in Chapter 2.4 are applicable to both loading and releasing of active ingredients. For the drug loading, the process is simpler. This is because there are many options for selecting a suitable solvent or solvent mixture in order to dissolve the drug and enable loading into an inorganic CRT system by a surface adsorption process. Molecular interaction between the drug/s and an inorganic CRT system is the important factor to be considered. This is because the solvent or the mixture of the solvents will affect the way a drug molecule will attach to the surface of a nanopore in an inorganic CRT matrix, which will then affect the enthalpy of ΔH_{ai} in the above equations and subsequently affect drug release from the inorganic CRT system.

8.2.3 Design and Development of an Inorganic CRT System for Drug Controlled Release

This book has introduced a technology platform based on inorganic materials and processes for delivering APIs (active pharmaceutical ingredients). By using a selection of synthesis methods for a wide range of inorganic materials, (also refer to Figure 2.14), the inorganic CRT material properties can be carefully modified. This allows a significant level of versatility in the choice of materials and subsequent control over the drug releasing property of the formulation such that the final inorganic CRT/drug composite can be carefully chosen to match the specific requirements of the drug to be delivered.[1] Inorganic CRT materials formulations can be designed and modified to cope with a range of various conditions to suit different applications. Key factors could be, but are not limited to

(a) chemical durability and solubility,
(b) drug loading and controlled release rate to maximize the efficacy,
(c) porosity (size, number) of nanostructure and
(d) an inorganic CRT which can be inert or active as required.

While inorganic CRT technologies are a relatively recent development for drug delivery, some common core approaches towards particular formulations for specific types of applications can already be highlighted.

It can be seen that Figure 8.1 represents a very general technology roadmap demonstrating which inorganic CRT materials are most particularly suited to which type of application. Following the reader's own critical evaluation of the formulation examples highlighted in this book, it should be clear that certain application areas and material types have been more extensively researched than others. For example, ordered silicas have been very well studied over recent years whilst clays and layered double hydroxide (LDH) materials have been significantly less well researched. As is the case for most new formulation types, enabling improved dissolution and bioavailability of poorly water-soluble compounds is by far the subject of most new formulation research studies – for good reason since this is where the area

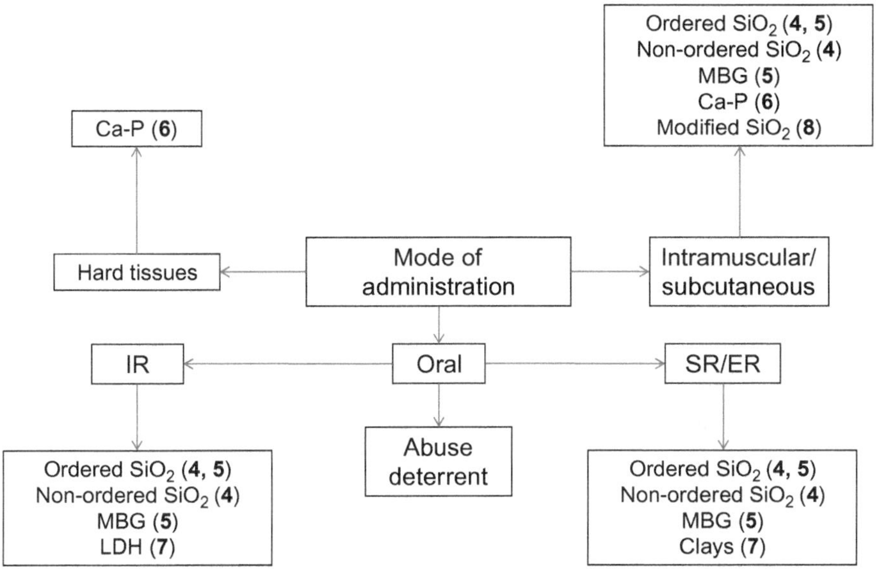

Figure 8.1 Overview of inorganic CRT approaches towards different formulations for different therapeutic areas. Numbers in parentheses refer to the appropriate chapter in this book. Ca-P, calcium phosphate ceramic; MBG, mesoporous bioactive glass; IR, immediate release; SR, sustained release; ER, extended release; LDH, layered double hydroxide.

of pharmaceutical formulation science has the most room for growth and development in terms of producing market-ready products. This should not mean, however, that there is no great value for carrying out new formulations research for the areas of very soluble drug compounds or for parenteral products for instance.

Figure 8.1 gives a general guide, but again, due to the relative immaturity of the field, the potential for matching new materials for applications not listed here, such as inhalables, is yet to be investigated on a significant enough scale to generate a substantial body of evidence. As has been described throughout this book, the potential for modification of the base carrier materials in order to modify drug release has been highlighted, but the limits of the individual technologies are still a long way from being defined, and in fact may never be so due to the inherent reliance on the physical and chemical properties of the individual therapeutic drug compounds. However the reader is invited, and encouraged, to take up research in this area and expedite the progression of learning in the hope that the full potential of this new field of controlled release technology will be realized.

In Chapter 1, we discussed the four classes of drugs according to the biopharmaceutics classification system, in which we particularly highlighted the best use of inorganic CRT for very soluble and poorly water-soluble drugs. The definition of a range of solubility is collected in Table 8.1 with seven solubility expressions. In this last chapter, we have no intention of discussing all of these different levels of drug solubility individually, but will highlight a few types of drugs as examples that are considered of importance for future inorganic CRT development.

8.2.3.1 Design and Development of Inorganic CRT Oral Formulations for Poorly Soluble Drugs

At present, many of the new potential therapeutics have poor pharmacokinetic and biopharmaceutical properties. Therefore, there is a need to develop suitable drug delivery systems that distribute the therapeutically active drug molecule in a beneficial manner, resulting in an ideal drug release profile. With this control, the inorganic CRT system can be used to increase solubility and bioavailability resulting in lower doses required for the required efficacy.

For the class of drugs that are labelled as practically insoluble drugs it is not possible to dissolve these types of drugs to concentrations that have any

Table 8.1 Solubility definition in the USP

Solubility definition	Parts of solvent required for one part of solute	Solubility range (mg/ml)	Solubility assigned (mg/ml)
Very soluble (VS)	<1	>1000	1000
Freely soluble (FS)	From 1 to 10	100-1000	100
Soluble	From 10 to 30	33-100	33
Sparingly soluble (SPS)	From 30 to 100	10-33	10
Slightly soluble (SS)	From 100 to 1000	1-10	1
Very slightly soluble (VSS)	From 1000 to 10,000	0.1-1	0.1
Practically insoluble (PI)	$>10,000$	<0.1	0.01

medical efficacy *in vivo*. The only way to dissolve a drug of this class is to design a drug carrier to deliver the drug from a more favourable thermodynamic position.

Strategy: A water-soluble inorganic CRT carrier with a modified nanopore surface should be developed to enhance the interaction between the poorly soluble drug molecules and the carrier surface function groups. Drug will be dispersed and distributed to a pseudomolecular level. A very rapidly dissolving inorganic CRT system will work well here such that the rate of absorption *in vivo* for these poorly soluble compounds will be enhanced as a result of drug super-saturation upon carrier dissolution. Drug molecules will be delivered depending on the dissolution rate of the inorganic CRT system.

8.2.3.2 Design and Development of Inorganic CRT Oral Formulations for Very Soluble Drugs

The preparation of immediate release formulations for very water-soluble compounds is a relatively straightforward process, and there are many formulations based on more traditional materials science which already perform this task well. In terms of controlling the active release and dissolution rate, there are fewer advantages of using inorganic CRT materials for this purpose. However, inorganic CRT materials as formulations do have other beneficial properties (see Chapter 1 for a full discussion) such as biocompatible degradation products and simpler processing methods.

The preparation of sustained release formulations for these types of drug compounds is significantly more challenging however. What is required is a system whereby the rate of drug dissolution from the carrier system is very much slower than the rate of diffusion of the solvated active molecules. Ways of achieving this include ensuring the carrier surface/drug interactions are strong and also by using carrier systems whereby the diffusion of solvated drug molecules out of the inorganic CRT system is sterically or electronically hindered.

Strategy: Templated and non-templated mesoporous silicas and mesoporous bioactive glasses would be suitable for this application due to the high number of surface functional groups and the potential for these groups for further modification. Their mesoporous interconnected pore structure, which results in sterically hindered drug diffusion out of the carrier, also makes these materials promising for sustained release formulations of highly soluble compounds. Certain clay materials would also be suitable provided that insoluble carrier matrices were employed under *in vivo* conditions. However, depending on the individual system used and the capacity for ion exchange under *in vivo* conditions, clay formulations as a general rule release their full payload over shorter time frames.

8.2.3.3 Design and Development of Inorganic CRT for Abuse-Deterrent Application

A particularly active area of research currently is the development of sustained release opioid formulations having abuse-deterrent properties. There are clear benefits to developing sustained release formulations for the management of chronic pain, but the necessity for incorporating high doses of the active constituents raises significant compliance issues. The presence of such high quantities in a single dosage form makes the patient more susceptible to addictive, dependent behaviour and the risk of respiratory depression due to a dangerously high *in vivo* active release is significant. The presence of high levels of the API also makes these formulations attractive for intentional abuse.

Therefore, the preparation of formulations demonstrating abuse-deterrent properties is now a requirement for future approval of these types of formulations.[2] It is recommended that these formulations are engineered to prevent unintentional dose dumping, intentional active extraction and increased drug release *via* crushing, melting, snorting, smoking, injecting and solvent extraction.

Strategy: It has recently been shown[3,4] that the use of opioid-loaded iCRT-deter materials results in differential release profiles in certain dissolution media. Active release has been shown to be significantly reduced in alcohol-based media as opposed to aqueous buffers. This dissolution property can be described as an abuse-deterrent facet since this behaviour satisfies the conditions of preventing accidental dose dumping under the influence of alcohol and intentional API extraction with appropriate solvents. In addition, the inherent properties of iCRT-deter materials necessarily means that they cannot easily be crushed, they do not melt at standard achievable temperatures and injection would result in serious complications due to the size of the available particles. These additional iCRT-deter formulation features all qualify as abuse-deterrent properties which makes these systems extremely promising for the development of sustained release, abuse-deterrent formulations.

8.2.3.4 Design and Development of Inorganic CRT for Protein Therapeutics

Most recently proteins have emerged as a major new class of pharmaceuticals.[5] Many proteins can serve as new classes of therapeutics, which have since been delineated into four distinct categories of therapeutic action; enzymatic/regulatory activity, targeting of specific molecules, vaccines and diagnostics.[6] Unfortunately the lability of proteins to hydrolytic and enzymatic cleavage during ingestion/digestion results in extensive protein degradation by first pass metabolism. This makes formulating proteins for oral delivery an exceptionally challenging task.

To avoid these metabolism effects and improve the bioavailability of protein therapeutics, formulation of proteins has focussed on administration routes that can bypass as many of the natural first pass metabolism processes as possible. These include intravenous, intramuscular, inhalation, transdermal and sublingual routes of administration during which the active protein is introduced directly into the systemic circulation.

As inorganic CRT drug delivery systems are such a new focus area of drug delivery research, most technologies have concentrated on the delivery and associated benefits of formulating small molecules for oral delivery (one exception is the use of drug-loaded calcium phosphates for the treatment of osseous defects, see Chapter 6). However, the perceived value and huge increase in interest for protein-based therapeutics for a range of conditions have led to an increase in efforts using new formulation technologies to help deliver these biomolecules. Inorganic CRT systems can provide effective solutions for the controlled delivery of protein therapeutics, hence we encourage more research to be carried out into this new therapeutic field to produce new families of protein-based drug delivery solutions.

Strategy: The preparation of porous silica systems by a modified method has shown great promise for the delivery of proteins. The formation of the silica polymeric network by an acid-catalysed sol-gel process occurs first followed by the addition of the relevant protein to a buffered solution before the sol-gel network has had chance to fully condense. This results in the semi-permanent encapsulation of the protein within the porous silica network. It has been demonstrated that encapsulation in this manner confers on the protein extra thermal and ionic stability whilst retaining a significant portion of the protein (enzymatic) activity. The encapsulation is only semi-permanent as these formulations are primarily being investigated as intramuscular implants and as such the protein, depending on its size and chemical potential, can be released by both diffusion and silica matrix dissolution-mediated processes.[7–9]

8.3 REGULATION AND RELEVANT ISSUES

8.3.1 Introduction

This section aims to provide basic knowledge on Quality System Regulation (QSR) for the readers who are not QSR professionals. QSR is the law. However, it varies from one country to another because every country has its own 'version' of QSR. The most well-known QSR is Good Manufacturing Practice, the main responsibility of which is assigned to the US Food & Drug Administration (FDA). FDA is an agency within the US Department of Health and Human Services, whose key responsibility is to protect the public health by assuring the safety, effectiveness, quality and security of healthcare products in the US market. QSR is a system that would assumingly lead the products made to meet these requirements. Here we introduce some concepts and practice as to what QSR is and where it fits into R&D and/or product development and will provide a simplified roadmap to follow in order to achieve compliance to worldwide regulations.

8.3.2 Design Control for Product Development

A design control is critical for compliance to QSR, which is especially applicable for research and development activities. So, what is design control? It is a systematic process with a range of activities that will guarantee the product development meets the regulation requirements, which include the following key elements:

(1) design and development planning,
(2) design input,
(3) design output,
(4) design review,
(5) design verification,
(6) design validation,
(7) design transfer,
(8) design changes,
(9) design history file (DHF).

In general, design control is about planning, design execution, testing against specifications (inputs) and requirements (user needs), and uncovering and fixing problems at an early stage (risk reduction). It is important that an organization gets used to planning and organizing team action and documentation for all the activities listed above from (1) to (8). Team action and documentation are highlighted in the design control flowchart in Figure 8.2 because of the critical part they play in obtaining compliance to QSR. For every project and/or subproject, it is good practice to have a planning, review and approval system, with good documentation, in place for the whole team. We will now move on to discuss the design control in detail.

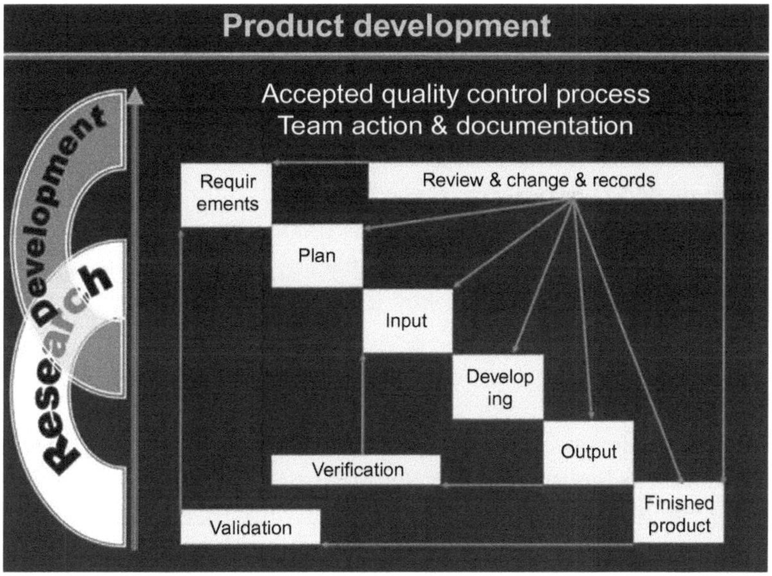

Figure 8.2 Design control.

8.3.2.1 Design and Development Planning

What should be in the plan? With regard to design control, a plan is a top level document and a full description of processes involved during product design and development, including roles, responsibilities and measurable deliverables. It documents planned actions such as peer reviews, major reviews and decision points. In general, a development plan is a plan that ensures that the design process is controlled and follows regulatory guidance. It should include, at least, three parts:

- Project plan: The project leader is responsible for creating a 'Project plan' for each project/subproject undertaken with tasks approved and resources assigned.
- Project schedule: To include planned actions and major milestones for the project. As design and development progresses, the 'Project plan' will be reviewed, updated and approved, where applicable; these changes also need to be reviewed, approved and documented.
- Project team: The Project Leader chooses a 'Project team' made up of members who are qualified to meet the needs of each assignment.

8.3.2.2 Design Input

Firstly, let us clarify the two concepts of 'user needs' and 'design inputs'. 'User needs' are the required functions from the customer's perspective (often the intended use and the intended use environments are defined by the market needs), while design input is the performance/requirements, including hardware, software and all other specifications, that make the required specifications suit the needs in design, development and manufacturing. From a design control point of view, user needs have to be validated independently while design inputs need verification by the design and development team to see if the 'inputs' meet or match the 'outputs'. *Validation* and *Verification* and fail or pass criteria should be detailed in the plan.

Design requirements are the key inputs to the design process. Design requirement specifications must document all of the characteristics that are essential for the proper functioning of the product; whenever needed and applicable; these are updated and approved as the design evolves. The detailed inputs and associated risk/potential risk analysis should be reviewed and approved throughout the development process using approved design control procedures. Changes to the product requirements must be reviewed and approved too in the same manner. Incomplete, ambiguous or conflicting requirements should also be identified and addressed *via* the design review process and then documented.

8.3.2.3 Design Output

In the simplest terms: Design output = the product + packaging and labelling + product master record.

Design outputs should conform to design input requirements, which should be structured so as to be traceable to that purpose. Design output procedures should contain or make reference to acceptance criteria. These procedures are there to ensure that the outputs are adequate enough to evaluate functions of a product, which may include product subsystems, outputs from risk analysis and final product specifications. It is necessary to specify and enumerate the design output as part of the roles and deliverables in the project plan, and then revise and update it whenever needed.

Deliverables of outputs should, at least, include

- product specifications,
- risk analysis,
- packaging and labelling and
- product master record (DMR).

8.3.2.4 Design Review

In general, formal design reviews are intended to

- assess design results, including designs for production and support processes;

- provide feedback to designers, such as problems/potential risks found;
- assess project progress and
- provide confirmation that the project is ready to move onto the next stage of development.

Review meeting and records (meeting minutes) should be well documented.

8.3.2.5 Design Verification/Validation

In this section, verification and validation (V&V) are discussed together because they follow the same methodology, albeit they have different goals.

Verification is confirmation by examination of objective evidence that the design output meets the functionality and performance specified by the design input. Verification activities should be focussed on answering the question: 'Did we design the product correctly?' That is in accordance with the correct requirements and with the right parts?

Validation is the process of evaluating the product at the end of the development process (or following a design change to existing products) to confirm that the product conforms to defined user needs and intended uses under actual or simulated use conditions.

If good requirements and specifications have been written, verification and validation are straightforward. Deliverables of V&V should include

- verification or validation plan,
- verification or validation protocol,
- verification or validation report and
- risk analysis and trace matrix of risks and potential risks.

Another question people often ask is how many manufacturing batches should be run for V&V? Regulatory authorities have not defined this number. There is a so-called magic 3, that is running three times for V&V for each of a planned verification and planned validation is probably the starting point, which is normally acceptable by regulation bodies. Here 'planned' means within design control, three runs are normally acceptable to the regulation authorities. However, it is the responsibility of the manufacturer or organization to decide the number of runs for verification and validation which they feel are adequate enough to meet the design specifications and requirements. V&V should be treated seriously and any variation, rather than simply confirmation of expected behaviour, should be sought. As well as doing it for compliance reasons, a manufacturer or responsible organization who performs V&V is also safeguarding their own interest in the long term.

8.3.2.6 Design Transfer

Design transfer is the correct translation of the product design into production specifications and the transfer of responsibility from design and development team to the manufacturing department to actually make the product. Transferring knowledge from the design team to the manufacturing and support functions should take place throughout the design process. Good industrial practice includes training, that is planned training to make sure that both the responsibility and the knowledge are transferred to the manufacturing. Deliverables of design transfer include

- a comprehensive design transfer plan which covers initial operation plan, final operation plan and process validation and
- DMR.

8.3.2.7 Design Change

There are two elements that must be taken into consideration: document control and change control. Document control includes establishing and maintaining procedures to control all design documents, including approval and distribution, the tracking of their status and revision history. Change control means enumeration of deficiencies and corrective actions arising from verification and review of the design, and tracking their resolution prior to design transfer. The question is again about responsibility. Before design transfer, design changes are the responsibility of the design team. After that, product and process changes are the responsibility of manufacturing.

8.3.2.8 Design History File

The DHF is a compilation of records which describes the design history of a finished product. They have been highlighted in the previous points and include

- design plan: project plan, schedule and design team;
- design inputs: design requirement documents and risk analysis;
- design outputs: product specifications, risk analysis, packaging and labelling, DMR and traceability files;
- design review: all review files;
- design verification: test plan, protocols and reports;
- design validation: test plan, protocols and reports;
- design transfer: DMR, process validation plans, protocols and reports;
- product release: authorization to ship and
- design change: all change files.

It is not intended to make this section a QSR training textbook but to highlight key elements that are important to the research and development of medical products in R&D environments to ensure full compliance with QSR. Without compliance, and therefore without regulatory approval, healthcare products simply cannot be introduced into the market. Achieving compliance must be an inherent and systematic approach across the whole organization, with planning, review, approval, documentation and management control standing as the central factors in that approach.

8.3.3 Note for Design Drug Development

The above QSR Design Control has highlighted the key regulatory requirements that are applicable for any healthcare products. Here we highlight a few important issues that are relevant to the design control for drug development in R&D environments.

8.3.3.1 Drug Substance

The physicochemical and biological properties of the drug substance that can influence the performance of the drug product and its manufacturability, or were specifically designed into the drug substance (e.g. solid-state properties), should be identified and discussed. Examples of physicochemical and biological properties that should be examined, as appropriate, include solubility, water content, particle size, crystal properties, biological activity and permeability. These properties could be interrelated and, when appropriate, should be considered in combination.

To evaluate the potential effect of drug substance physicochemical properties on the performance of the drug product, studies on drug product might be warranted. For example, ICH Q6A Specifications: Test Procedures and Acceptance Criteria for New Drug Substances and New Drug Products: Chemical Substances describes some of the circumstances in which drug product studies are recommended (e.g. Decision Trees #3 and #4 (Part 2)). This approach applies equally for ICH Q6B Specifications: Test Procedures and Acceptance Criteria for Biotechnology/Biological Products. The knowledge gained from the studies investigating the potential effect of drug substance properties on drug product performance can be used, as appropriate, to justify elements of the drug substance specification (3.2.S.4.5).

8.3.3.2 Excipients

The excipients chosen, their concentration and the characteristics that can influence the drug product performance (e.g. stability, bioavailability) or manufacturability should be discussed relative to the respective function of each excipient. This should include all substances used in the manufacture of the drug product, whether they appear in the finished product or not (e.g. processing aids). Compatibility of excipients with other excipients, where relevant (e.g. combination of preservatives in a dual-preservative system), should be established. The ability of excipients (e.g. antioxidants, penetration enhancers, disintegrants, release controlling agents) to provide their intended functionality and to perform throughout the intended drug product shelf life should also be demonstrated. The information on excipient performance can be used, as appropriate, to justify the choice and quality attributes of the excipient and to support the justification of the drug product specification.

8.3.3.3 Formulation Development

A summary should be provided describing the development of the formulation, including identification of those attributes that are critical to the quality of the drug product, taking into consideration intended usage and route of administration. Information from formal experimental designs can be useful in identifying critical or interacting variables that might be important to ensure the quality of the drug product. The summary should highlight the evolution of the formulation design from initial concept up to the final design. This summary should also take into consideration the choice of drug product components (e.g. the properties of the drug substance, excipients, container closure system, any relevant dosing device, the manufacturing process, and, if appropriate, knowledge gained from the development of similar drug product(s)).

A summary of formulations used in clinical safety and efficacy and in any relevant bioavailability or bioequivalence studies should be provided. Information from comparative *in vitro* studies (e.g. dissolution) or comparative *in vivo* studies (e.g. bioequivalence) that links clinical formulations to the proposed commercial formulation should be summarized and a cross-reference to the studies (with study numbers) should be provided. Where attempts have been made to establish an *in vitro/in vivo* correlation, the results of those studies and a cross-reference to the studies (with study numbers) should be provided in the drug development document. A successful correlation can assist in the selection of appropriate dissolution acceptance criteria and can potentially reduce the need for further bioequivalence studies following changes to the product or its manufacturing process.

8.3.3.4 Physicochemical and Biological Properties

The physicochemical and biological properties relevant to the safety, performance, or manufacturability of the drug product should be identified and discussed. This includes the physiological implications of drug substance and formulation attributes. Specifications of regulatory requirements are covered in two references '*ICH Q6A Specifications: Test Procedures and Acceptance Criteria for New Drug Substances and New Drug Products: Chemical Substances, Decision Tree #4 (Part 3) and Decision Tree #7 (Part 1)*' or '*ICH Q6B Specifications: Test Procedures and Acceptance Criteria for Biotechnology/Biological Products*'.

8.3.3.5 Microbiological Attributes

Where appropriate, the microbiological attributes of the drug product should be discussed and reported. Although chemical testing for preservative content is the attribute normally included in the drug product specification, antimicrobial preservative effectiveness should be demonstrated during development. The lowest specified concentration of an antimicrobial preservative should be demonstrated to be effective in controlling microorganisms by using an antimicrobial preservative effectiveness test. The concentration used should be justified in terms of efficacy and safety, such that the minimum concentration of preservative that gives the required level of efficacy throughout the intended shelf life of the product is used. Where relevant, microbial challenge testing under testing conditions that, as far as possible, simulate patient use should be performed during development and documented.

8.3.3.6 Compatibility

The compatibility of the drug product with reconstitution diluents (e.g. precipitation, stability) should be addressed to provide appropriate and supportive information on labelling. This information should cover the recommended in-use shelf life, at the recommended storage temperature and at the likely extremes of concentration. Similarly, admixture or dilution of products prior to administration (e.g. product added to large volume infusion containers) should be addressed.

8.4 SUMMARY

The aim of pharmaceutical development is to design a quality product and its manufacturing process to consistently deliver the intended performance of the product. The information and knowledge gained from pharmaceutical development studies and manufacturing experience provide scientific understanding to support the establishment of the design space, specifications and manufacturing controls.

REFERENCES

1. Campbell I. Lucideon Ltd. Towards the future of controlled release applications, www.lucideon.com [accessed 02.04.15].
2. Food Drug Administration (FDA), Center for Drugs Evaluation Research (CDER). *Guidance for industry: abuse-deterrent opioids – evaluation and labelling*. MD: FDA; 2015.
3. Lucideon patent pending. 2014.
4. Lucideon patent pending. 2015.
5. Walsh G. Biopharmaceutical benchmarks 2010. *Nat Biotechnol* 2010;**28**:917–24.
6. Leader B, Baca QJ, Golan DE. Protein therapeutics: a summary and pharmacological classification. *Nat Rev Drug Discov* 2008;**7**:21–39.
7. Avnir D, Braun S, Lev O, Ottolenghi M. Enzymes and other proteins entrapped in sol-gel materials. *Chem Mater* 1994;**6**:1605–14.
8. Teoli D, Parisi L, Realdon N, Guglielmi M, Rosato A, Morpurgo M. Wet sol-gel derived silica for controlled release of proteins. *J Control Release* 2006;**116**:295–303.
9. Santos EM, Radin S, Ducheyne P. Sol-gel derived carrier for the controlled release of proteins. *Biomaterials* 1999;**20**:1695–700.

FURTHER READING

Guidance for Industry Q8(R2) Pharmaceutical Development', US Department of Health and Human Services Food and Drug Administration Centre for Drug Evaluation and Research (CDER) Centre for Biologics Evaluation and Research (CBER) November 2009 ICH Revision 2.

ISO 13485:2003 Medical Products – Quality Management Systems.

ISO 9001: subclause 4.4.

USA Food and Drug Administration, 21 CFR Part 820.30, Quality System Regulation.

USA Food and Drug Administration, 21 CFR Part 820, USA Food and Drug Administration, Quality System Regulation, 1996.

INDEX

Note: Page numbers followed by *f* indicate figures and *t* indicate tables.

A

Abuse-deterrent, 232
Acetylcholinesterase inhibitor, 33, 34*t*
Acne treatment, 131
Adenosine triphosphate (ATP), 161, 163*f*
Aerosil®, 12–14
α-tricalcium phosphate (α-TCP), 166
Alumina, 208–210, 209*f*
Amorphous materials
  free volume theory, 22–23, 23*f*
  glass transition, 21, 22*f*
  history of, 18
  sol-gel processing, 94–95
  structure of, 18–21
Amphotericin B, 196
Angiogenic growth factors, 152–153
Anionic clays. *See* Layered double hydroxides (LDHs)
Ansamycins, 33, 34*t*
API (active pharmaceutical ingredient)
  aspirin distribution, 24–25, 24*f*
  dissolution rate, 28–29
  drug-loaded mesoporous silica, 27–28, 28*f*
  genotoxic impurities, 86, 87*f*
  ibuprofen-loaded sol-gel glass particles, 26–27, 27*f*
  and iCRT matrix interaction, 29–30
  multi API distribution, Parkinson's disease, 24–25, 24*f*
  nano-dispersion and distribution, 29
  particle size reduction, 28–29
  and P-glasses interaction, 43–44, 44*f*
  purity, separation and detection methods, 85–86, 85*t*
  tablet processing flow chart, 25, 25*f*
  toxicity evaluation (*see* Toxicology)
Aspirin
  chemical structure, 33, 34*t*
  distribution, 24–25, 24*f*
  molecular structure of, 24–25, 24*f*
Atenolol, 119

B

Backscattered electrons (BSEs), 82–83
Benserazide, 24–25, 24*f*
Benzoyl peroxide
  acne treatment, 131
  benefits of, 133
  encapsulated porous microspheres, 133, 133*f*
  formulations, 133–134
  rosacea treatment, 131
β-tricalcium phosphate (β-TCP), 163
BET model. *See* Brunauer, Emmett and Teller (BET) model
Bioceramics, 12, 13*f*
Biodegradable polymers, 1
Bioglass®, 94
Block copolymer template
  Flory-Huggins theory (*see* Flory-Huggins theory)
  Hildebrand solubility parameter, 99–101
  ordered polymer templates, 104, 106–111
  solvent evaporation approach, 110–112, 112–113*f*
Bone tissue engineering (BTE), 139–140, 146
Bragg's law, 72–73, 73*f*
Brunauer, Emmett and Teller (BET) model
  adsorption energy, 79
  equilibrium and saturation pressures, 80
  molecular adsorption, 79
  saturation condition, 80
  silica MCM-41, nitrogen adsorption, 80–82, 81*f*
  slope and intercept, 80
BSEs. *See* Backscattered electrons (BSEs)
BTE. *See* Bone tissue engineering (BTE)
Burst effect, 3
Butorphanol, 33, 34*t*

C

Calcium deficient hydroxyapatite (CDHA), 163
Calcium phosphate
  ATP, hydrolysis of, 161, 163*f*

Calcium phosphate *(Continued)*
biological functions of, 161
cements (*see* Calcium phosphate cements)
coatings, 179–183, 182–183*f*
dental hygiene products, 164
foams, 178–179, 180–181*f*
HA and natural bone, compositions of, 163, 163*t*
lab-produced bone void, 164
properties of, 162*t*, 163
Calcium phosphate cements
apatite-forming reactions, 166
benefits of, 164
bioresorption rate, 164–165
brushite-forming reactions, 166
in clinical setting, 165–166
controlled drug release rates, 171–176, 175–176*f*
crystal interlocking process, 167
dissolution kinetics, 167–168
as drug delivery systems, 169–170
drug-loading methods, 170–171, 171*f*
indomethacin, 176–177
liquid:powder ratios, 168, 169*f*
macroporous network, 167, 168*f*
microporous voids, 167
new crystal phases, formation of, 166–167
plate-like crystals, 167–168, 169*f*
vs. PMMA cement setting temperatures, 164, 165*t*
porous implants, 164
RANKL, 177
setting process, 165–167
strontium, 177, 178*f*
surface reactivity, 167–168
Camptothecin, 194–196, 195*f*
Cancer therapy, 29
Carbacephems, 33, 34*t*
Cationic clays
chlorites clay, 203, 203*t*
illites clay, 203, 203*t*
kaolinite clays, 203, 203*t*, 206
natural clay systems, 202–203
smectite clays, 203–206, 203*t*
tetrahedral-octahedral arrangement, 201–202, 202*f*
vermiculite clay, 203, 203*t*, 206–207
Cerium (Ce^{3+}), 151
Chemical analysis
FTIR spectroscopy, 63–66
ICP-MS (*see* Inductively coupled plasma mass spectrometry (ICP-MS))
SIMS technique, 68–72
XPS technology, 66–68
X-ray fluorescence, 59–60, 60*t*
Chlorites clay, 203, 203*t*
Citric acid, 24–25, 24*f*
Clays
anionic (*see* Layered double hydroxides (LDHs))
cationic (*see* Cationic clays)
ion exchange, 190
2:1 natural clay, 190, 191*f*
talc, 190
Cocondensation method, 146–148
Controlled release technology (CRT)
biodegradable polymers, 1
definition, 1
inorganic materials (*see* Inorganic controlled release technology (iCRT))
and non-controlled drug release profiles, 1, 2*f*
polymeric organic CRTs (*see* Organic controlled release technology)
slow/sustained controlled release, 1–2
triggered release, 1–2
Cyclandelate, 33, 34*t*

D

Design and development, iCRT system
abuse-deterrent, 232
compatibility, 241
drug substance, 239
excipients, 239–240
formulations, 228–229, 229*f*, 240
highly soluble drugs, 231–232
microbiological attributes, 241
physicochemical and biological properties, 241
poorly soluble drugs, 230–231
protein therapeutics, 232–233
QSR (*see* Quality System Regulation (QSR))
solubility definition, 230, 230*t*

Design control
definition, 234
design and development planning, 234–235
design change, 234, 238
design history file, 234, 238–239
design transfer, 234, 238
flowchart, 234, 235*f*
inputs, 234–236
output, 234, 236
reviews, 234, 236–237
user needs, 235–236
verification and validation (V&V), 234, 237
Design history file (DHF), 238–239
Dexamethasone (DEX), 151
Dimethyloxalylglycine (DMOG), 151
Dipyridamole, 33, 34*t*
Donepezil, 204–205, 204–205*t*
Doxorubicin, 213

E

Elastic scattering, 72–73, 82–83
Energy dispersed X-ray (EDX) detector, 83
Eudragit® E-100 copolymer, 200–201, 205
Evaporation-induced self-assembly (EISA) process, 141, 142*f*
Excipients, 88, 239–240
and API, in drug formulations, 29–30
known excipient, 88–90
new excipient, 87–89

F

Fickian diffusion model, 48–53
Flory-Huggins theory
binary block copolymer, 102, 102*f*
Gibbs free energy, 101–102, 105*f*
interaction parameters χ_{AB}, 102–104
interaction parameters $N\chi_{AB}$, 104–106, 105*f*
mixing entropy change, 102
polymer blends, assessment of, 101
Food & Drug Administration (FDA), 233–234
Fourier transform infrared (FTIR), 59
advantages of, 63
limitation of, 66
for quantitative analysis, 63
silanol groups, 64–65, 64*f*
SiO—NB/SiO—B, 65–66, 66*f*
working principle, 63–64
Free volume theory, 22–23, 23*f*
Fumed silica, 128
Fusion glasses
constituents and functions, 7, 8*t*
drug-loaded water-soluble glasses, 7, 9*f*
oxides, high temperature processing of, 7, 8*f*
traditional preparation, 7

G

Gallium (Ga^{3+}), 151
Gentamicin sulphate (GS), 152, 153*f*
Gibbs free energy, 101–102, 105*f*
Glass transition theory, 21, 22*f*
Gold nanoparticles, 214–216, 215*f*
Gramicidin, 196

H

Halloysite, 206, 207*f*
hBMSCs. *See* Human bone marrow stromal cells (hBMSCs)
Hildebrand solubility parameter
cohesive energy density, 99–100
hydrogen bonding, 100–101
London dispersion forces, 100
measurement and estimation, 101
mixing enthalpy, 100
mixture volume, 100
non-polar materials, 100
polar cohesive forces, 100
volume fraction, 100
Human bone marrow stromal cells (hBMSCs), 149–150
Human periodontal ligament cells (hPDLCs), 149–150
Hydrotalcites, 190
Hypoxia, 149–150

I

Ibuprofen
chemical structure, 33, 34*t*
logarithmic time, 52, 52*f*

Ibuprofen *(Continued)*
and mesoporous silica MCM-41, 33–38, 36*f*, 52, 52*t*, 118–119, 119*f*
sol-gel glass particles, 26–27, 27*f*
water-soluble glass, controlled release from, 46, 46*f*
iCRT. *See* Inorganic controlled release technology (iCRT)
Illites clay, 203, 203*t*
Implant delivery dissolution model, 125–126, 126*f*
Indomethacin, 125–126, 125–126*f*, 176–177
Inductively coupled plasma mass spectrometry (ICP-MS)
advantage of, 60–61
cobalt and chromium, metal-on-metal implants, 61–63
controlled strontium release, P-glass, 61, 62*f*
limitation, 63
working principle, 61
Inelastic scattering, 82–83
Inorganic controlled release technology (iCRT)
API (*see* API)
calcium phosphate (*see* Calcium phosphate)
chemical analysis (*see* Chemical analysis)
clays (*see* Clays)
commercial application of, 4
design and development (*see* Design and development, iCRT system)
drug-loaded micro/nanostructure, 12–15
fusion glasses, 7–8
kinetic theories, 226–228
Lucideon iCRT, 4
MBGs (*see* Mesoporous bioactive glasses (MBGs))
mesoporous alumina, 208–210, 209*f*, 211*f*
mesoporous silicas (*see* Mesoporous silicas)
mesoporous zirconia, 207–208, 208*f*
molecular interactions (*see* Molecular interactions, iCRT systems)
nanoparticles (*see* Nanoparticles)
nanoporosity characterization (*see* Nanoporosity)
nanotechnology, definition of, 222
vs. polymeric organic CRT, 4–7, 5*t*, 6*f*
release kinetics, 48–53
sol-gel glasses (*see* Sol-gel silica glasses)
thermodynamic theories, 222–226
XRD technology (*see* X-ray diffraction (XRD))
Ionic surfactant template
Atenolol release, 119, 120*f*
binary block copolymers, 117–118, 118*f*
limitation, 119–120
liquid-crystal template mechanism, 117–118, 118*f*
mesoporous M41S materials, 117–119, 118–119*f*
synthesis conditions, 117
Iron oxide nanoparticles (IONPs), 211–213, 213*f*
Isoxsuprine, 33, 34*t*

K

Kaolinite clays, 203, 203*t*, 206

L

Layered double hydroxides (LDHs), 89
anionic exchange, 192–193, 192*f*, 193*t*
controlled drug delivery, 198–201, 200–201*f*
coprecipitation, 193–194
non-ionic compounds, 194–197, 195*f*
property, 191–192
reconstruction, 194, 194*f*
structure of, 191–192, 191*f*
Levodopa, 24–25, 24*f*
Lincosamides, 33, 34*t*
London dispersion forces, 100
Lucideon iCRT, 4

M

MBGs. *See* Mesoporous bioactive glasses (MBGs)
MC. *See* Methylcellulose (MC)
Memantine, 33, 34*t*
Mesoporous alumina, 208–210, 209*f*, 211*f*
Mesoporous bioactive glasses (MBGs)
biomedical applications, 139–140
bone defects, treatment of, 148–149

drug delivery and growth factor, 151, 152–153*f*
EISA method, 141, 142*f*
growth factors, 152–153
polymer coating, drug delivery from, 153–154
sol-gel process, 140–142, 141*f*
surface functionalization, 146–148, 147*f*
therapeutic ions, delivery of, 149–151, 149*t*
3D scaffold structure, 141–146, 142–143*f*, 145*f*
Mesoporous silicas
block copolymer template (*see* Block copolymer template)
dehydroxylation and rehydroxylation, 32, 33*f*
drug-loaded particles, 12–15, 14*f*, 27–28, 28*f*
ibuprofen and MCM-41, 33–38, 36*f*
ionic surfactant template, 117–120
low-molecular-weight surfactants, 112–113, 113–114*t*
non-ionic surfactant template, 114–116
properties of, 98
surfactant template approach, 10–12, 11*f*
Mesoporous zirconia, 207–208, 208*f*
Metal-on-metal hip implants, 61–63
Methotrexate, 198–199, 198*f*
Methylcellulose (MC), 143
Miller indices, 74, 74*f*
Molecular interactions, iCRT systems
API and iCRT matrix systems, 30–32
API excipients, in drug formulations, 29–30
$CaO-SiO_2-P_2O_5$ glasses, formulation roadmap, 47–48, 47*f*
drug molecules, 33, 34*t*
mesoporous silica MCM-41 and ibuprofen, 33–38, 36*f*
molecular dispersion and distribution, 38–39
P-glasses (*see* Phosphate glasses (P-glasses))
silica surfaces, chemistry of, 32, 33*f*, 38
sol-gel silica glasses, 39–40
Monobactams, 33, 34*t*
Montmorillonite, 204, 204–205*t*

N

Nanoparticles
gold nanoparticles, 214–216, 215*f*
IONP, 211–213, 213*f*
mesoporous silicas, 12–15, 14*f*
Nanoporosity
BET model (*see* Brunauer, Emmett and Teller (BET) model)
nitrogen adsorption, mesoporous material, 75–79, 78*f*
Nitrofurans, 33, 34*t*
4-Nitrophenol, 86, 87*f*
Non-ionic surfactant template, 114–116
Non-ordered porous silica
matrix dissolution, 130–131
processing of, 128–131
silicon alkoxides, sol-gel reactions of, 120–121, 127
sodium silicate, precipitation reactions of (*see* Precipitated silica)
N-(phenoxycarbonyl)-L-valine, 86, 87*f*

O

Oleate, 196
One-dimensional nanopore model, 49, 49*f*
Oral delivery dissolution model, 125–126, 126*f*
Ordered mesoporous silicates
block copolymer template (*see* Block copolymer template)
drawbacks, 120
ionic surfactant template, 117–120
low-molecular-weight surfactants, 112–113, 113–114*t*
non-ionic surfactant template, 114–116
properties of, 98
Organic controlled release technology
advantage of, 2–3
burst effect, 3
commercial products, 4
feature of, 3–4
vs. inorganic CRT, 4–7, 5*t*, 6*f*
variations and variables, 3–4
Osteogenic growth factors, 152–153
2-Oxazolidone, 33, 34*t*

P

Papulopustular rosacea, 131
Parkinson's disease, 24–25, 24*f*
Penicillin, 33, 34*t*
Phenol, 86, 87*f*
Phosphate glasses (P-glasses)
  for biomedical applications, 40
  characteristics, 40
  composition evaluation, XRF, 60, 60*t*
  controlled strontium release, 61, 62*f*
  dissolution rate, 40, 45–47
  oxides, cationic radius and normalized charges, 40–44, 41*t*, 42*f*, 44*f*
  surface chemistry of, 43–45, 44*f*
Physical characterization, iCRT materials
  nanoporosity, 75–82
  XRD technology (*see* X-ray diffraction (XRD))
Poly(tertbutyl acrylate-*co*-ethyl acrylate-*co*-methacrylic acid) (PTBEM), 196
Polymer CRT systems. *See* Organic controlled release technology
Porous silica
  benzoyl peroxide (*see* Benzoyl peroxide)
  non-ordered silica (*see* Non-ordered porous silica)
  oral and parenteral treatments, 134
  poorly water-soluble drug compounds, 131, 132*t*
  topical formulation, development of, 134
Precipitated silica, 121
  average pore size, 122
  diatom pores, SEM images of, 125, 125*f*
  experimental set-up, 122, 122*f*
  indomethacin, release of, 125–127, 126*f*
  natural resources, recycle and reuse, 123
  particle size and morphology, change in, 122–123, 123*f*
  pure diatoms, diatomaceous earth, 124, 125*f*
  silica gels, ageing and drying processes, 123–124, 124*t*
  surface area, 122
Prednisone, 196, 197*f*
Protein therapeutics, 232–233
PTBEM. *See* Poly(tertbutyl acrylate-*co*-ethyl acrylate-*co*-methacrylic acid) (PTBEM)

Q

Quality System Regulation (QSR)
  design control (*see* Design control)
  FDA, 233–234
  Good Manufacturing Practice, 233–234

R

Replication technique, 144
Ritonavir, 86, 87*f*

S

Scanning electron microscopy (SEM), 58
  aspirin tablet, 24–25, 24*f*
  BSE, 82–83
  diatom pores, 125, 125*f*
  EDX, 83
  ibuprofen-loaded sol-gel glass particles, 26–27, 27*f*
  pure diatoms, diatomaceous earth, 124, 125*f*
  secondary electron (SE), 82
Secondary electron (SE), 82
Secondary ion mass spectrometry (SIMS)
  bioactive glass, ToF-SIMS analysis of, 69–72, 70*f*
  dynamic and static SIMS, 68–69
  limitations, 72
  working principle, 69, 69*f*
SEM. *See* Scanning electron microscopy (SEM)
Silica precipitation process. *See* Precipitated silica
Sirolimus, 33, 34*t*
Smectites clay, 203–206, 203*t*
Sol-gel silica glasses
  amorphous XRD profile, 19–20, 20*f*
  chemical reactions for, 9, 10*f*
  constituents, 10, 11*t*
  drug-loading and final powder processing, 9–10, 10*f*
  free volume, 23, 23*f*
  low temperatures, 9
  molecular interactions, 39–40
  silicon alkoxides materials, 94–98
Solvent deposition method, 128–129
Solvent evaporation approach, 110–112, 112–113*f*

Strontium (Sr), 149–150
  calcium phosphate cements, 177, 178*f*
  P-glass, controlled release from, 61, 62*f*
Superparamagnetic iron oxide, 212

T

TEM. *See* Transmission electron microscopy (TEM)
Tetracalcium phosphate (TTCP), 166
Tetraethyl orthosilicate (TEOS), 9, 10*f*
3D crystal unit cell model, 74, 74*f*
Three-dimensional (3D) plotting technique, 145
Three-dimensional (3D) printing technique, 145–146
Time-of-flight secondary ion mass spectrometry (ToF-SIMS), 59
  bioactive glass, 69–72, 70*f*
  phosphate glasses, 43–44, 44*f*
Toxicology
  an active material, 88–89
  known excipient, 88–90
  new excipient, 87–89
Transforming growth factor-β1 (TGF-β1), 129–130
Transmission electron microscopy (TEM)
  drug-loaded mesoporous silica, 12–14, 14*f*, 27–28, 28*f*, 84
  high resolution, 83–84
  limitation, 84
  wavelength of, 83

V

Vancomycin, 175*f*
Vascular endothelial growth factor (VEGF), 152–153
Vermiculite clay, 203, 203*t*, 206–207
Vitamin E, 33, 34*t*

X

XPS. *See* X-ray photoelectron spectroscopy (XPS)
X-ray crystallography, 72
X-ray diffraction (XRD)
  bulk sample analysis, 58
  calcium hydroxyapatite (HA), 74–75, 76*t*
  crystalline and amorphous materials, 19–21, 20*f*, 72–75, 73–74*f*, 77*f*
  limitations of, 75
X-ray fluorescence (XRF)
  bulk sample analysis, 58
  elemental analysis, 59
  limitation, 60
  P-glass composition evaluation, 60, 60*t*
  working principle, 59
X-ray photoelectron spectroscopy (XPS), 59
  limitation, 68
  photo-ionization, 66
  Si—O—Si, SiO_2 content, 67–68, 68*f*
  working principles, 66–67

Z

Zirconia, 207–208, 208*f*

Kolibri 360 GmbH
Ferdinand-Jühlke-Straße 7
99095 Erfurt, Deutschland
produktsicherheit@kolibri360.de